To Kieran,

bright future

for a

with my compliments

07/12/23

Hardwiring Sustainability into Financial Mathematics

Armen V. Papazian

Hardwiring Sustainability into Financial Mathematics

Implications for Money Mechanics

Armen V. Papazian
King's College Cambridge
University of Cambridge
Cambridge, UK

ISBN 978-3-031-45688-6   ISBN 978-3-031-45689-3   (eBook)
https://doi.org/10.1007/978-3-031-45689-3

Cover illustration: © John Rawsterne/patternhead.com

This Palgrave Macmillan imprint is published by the registered company Springer Nature
Switzerland AG
The registered company address is: Gewerbestrasse 11, 6330 Cham, Switzerland

Paper in this product is recyclable.

For my future grandchildren…

FOREWORD

Our understanding of Finance and Investment is lacking an essential component, namely, its interaction with the rest of the world. Finance and Investment do not stand in splendid isolation and yet this is how we have long chosen to model our understanding of the dynamics at play between all the different actors and components involved. This gap is precisely what Dr. Armen Papazian bridged in his book *The Space Value of Money*. The clue was in the subtitle—he promised *Rethinking Finance Beyond Risk and Time*, a notion that is long overdue, and duly delivered.

In this book, Dr. Papazian goes a step further, this time discussing sustainability in financial mathematics, drawing once again on his concept of the space value of money. Although significant strides have already been made to properly account for our environmental and social impacts, this book's contribution is to advance our thinking with respect to the intricacies of accounting frameworks and standards, as well as money mechanics. Much remains to be done.

No one else has attempted to properly include concepts of sustainability in finance and investing in the way that the author does. This book should serve as the perfect tool to reformat our understanding of the topic and also to educate those that come after us.

Finance has a critical role to play and embedding it within the twenty-first century requires that we broaden our horizons—this is what this book allows the reader to achieve.

UK
September 2023

Dr. Saker Nusseibeh, CBE
Chief Executive Officer
Federated Hermes Limited

PREFACE

The idea of this book was born after a paper presentation at the European Central Banks' event organised by Palgrave Macmillan and Springer Nature in March 2023. While the main arguments and equations I present and discuss are based on my previous 2022 publication *The Space Value of Money: Rethinking Finance Beyond Risk and Time*, this manuscript differs in scope and approach. It is focused on making the case for the hardwiring of sustainability into financial mathematics, specifically our equations of value and return in finance, and the implications of such a transformation for money mechanics, the process of money creation and monetary policy transmission.

This book, like its predecessor, is a theoretical treatise on sustainability in finance. It does not seek to find correlations in the past, nor does it attempt to find statistical evidence to support its propositions. This is so for two key reasons, first, because it aims to offer a new value framework and equations that can allow us to change trajectory and therefore create a future very different from the past where potential statistical evidence may be found, and second, because only after the adoption and widespread application of the proposed equations can statistical back testing become potentially possible.

The raging climate crisis and its many and widespread manifestations, from droughts to heatwaves, wildfires, floods, and extreme hailstorms, have now become daily front page news. Rising emissions and CO_2 levels in our atmosphere, by the 4th of August 2023 at 421.23 parts per million, have been time and again found to be responsible for the warming of our planet and the disruptive climate events experienced across the world.

Our challenge, however, goes far beyond emissions and carbon. We have littered every environment we have come to touch—from oceans to land, rivers and lakes, atmosphere and outer space. Human productivity is evidently oblivious to and unconcerned with what it leaves behind, and how it endangers its very own continuity and evolution. A hit and run approach that looks like a heist on future generations, we seem more convinced and in service of our self-inflicted debt-based monetary architecture than the ecosystem that provides our sustenance and survival.

The recent rise to prominence of sustainable finance (SF), or ESG integration, or climate finance, or impact investing, is a testimony to a growing awareness of a fundamental imbalance in human productivity, and evidence of a concerted effort to redirect and guide it towards a more sustainable future. Indeed, the diverse initiatives aimed at conceptualising and operationalising sustainability in business, industry, and finance are ongoing, a work in progress with evolutionary implications.

Very recently, the International Sustainability Standards Board (ISSB) of the International Financial Reporting Standards Foundation (IFRS) released its inaugural standards. They were soon after endorsed by, amongst others, the Financial Stability Board (FSB) and the International Organisation of Securities Commissions (IOSCO). These developments indicate growing momentum towards a future where these standards will become mandatory. On the 31st of July 2023, the European Commission adopted the European Sustainability Reporting Standards (ESRS) to be used under the Corporate Sustainability Reporting Directive (CSRD), which are themselves aligned with the IFRS standards. It is therefore critical to ensure that these standards are effective and can actually shift our trajectory and lead us to a sustainable future.

Indeed, the importance of their effectiveness cannot be overstated as the implications of adopting ineffective frameworks and standards can have far-reaching consequences from an existential perspective, and a damaging impact on business and industry from an efficiency perspective. This is why, in fact, we have seen significant backlash against ESG. Greenwashing does not achieve the sustainability we seek, and the same goes for ESG integration in its current form.

The popularisation of ineffective sustainability 'non-solutions' that add new regulatory market requirements may create a whole new labour market niche, may expand the revenue margins of ESG ratings providers, consultants, and software providers, may increase the commissions of ESG peddling salespersons, but it will fall short of achieving the change we need. Given a rampant tendency to safe play, in industry and academia, these non-solutions can also hinder the search and adoption of the effective long-term solutions we need, which, more often than not, imply and require radical transformations in us, in our values, practices, models, and beliefs.

As I write this preface, we continue to consume our very own and only home from under our own feet, while expanding the checklist of regulatory and market requirements in an already challenging and unequal post-pandemic global context. While I support effective regulation, half measures will make business and industry suffer, without reaping the benefit of achieving the purpose, i.e., securing a healthier planet and a sustainable human civilisation.

The relentless upward trend in emissions, in pollution and waste, in biodiversity loss, and environmental degradation must be addressed at the source, i.e., money, the value framework that defines its value, and the equations that govern its creation, allocation, and deployment—taught and applied by many millions around the world.

To truly redirect human productivity, we must reimagine the value of money and the institutional structures that create it. Money and our monetary architecture are at the heart of the sustainability challenge and opportunity.

As disappointed as one may be by humanity's current failings, we must steer clear of a deeper hopelessness driven by a bleak interpretation of human nature. While such interpretations may be justified by the wars, violence, abuse, and injustice that seem to be rampant in our world today, against fellow humans, other species, and our entire ecosystem, they must be qualified by the love, kindness, and goodness that make this world go round.

Without hope or hopelessness, our predicament can only be addressed through our creative imagination and our will and resolve to seek and implement the appropriate transformations that could secure the sustainability and creative expansion of human productivity across time and space.

UK   Armen V. Papazian
August 2023

ACKNOWLEDGEMENTS

Like any project and endeavour of this kind, I owe thanks and gratitude to a number of institutions and individuals, who have, through their direct or indirect support, and contributions, made this book a reality.

I would like to thank:

Springer Nature, for inviting me to present at the European Central Banks' event in March 2023, a unique opportunity that has led to the eventual publication of this book.

Palgrave Macmillan, for continuously supporting my work and facilitating its worldwide distribution, a brilliant team that I have the pleasure to work with.

King's College Cambridge University, for continuously inspiring my work and hosting the launch of *The Space Value of Money* book, a memorable moment in my life as well as the history of the ideas I propose.

Judge Business School Cambridge University, for supporting my work, providing a platform for the ideas discussed in this book, and partnering in the launch of *The Space Value of Money* book.

Federated Hermes Limited, for providing research funding during the early stages of writing of *The Space Value of Money* book, the foundational work upon which this manuscript is based.

Space Value Foundation, for supporting and providing me with a unique platform to advocate for the ideas discussed in this book.

I owe special thanks to Tula Weis for her continuous support, for recognising the relevance of my work and making this publication possible, to Susan Westendorf, Ashika Joycell, and Karthika Purushothaman for the highly professional and efficient production that ensured the timely release of this book, and to the anonymous referees who provided valuable feedback and suggestions that improved this manuscript.

I owe special thanks to Dr. Saker Nusseibeh, CBE, for accepting the invitation to write a foreword, I am grateful for and inspired by his kind and encouraging words, to Eoin Murray, George Littlejohn, Dr. Matteo Cominetta, and Adrian Webb for their support, for reviewing the early drafts of this book, for providing reviews.

I owe special thanks to Dr. Keith Carne, Prof. Gishan Dissanaike, Dr. Pascal Blanqué, Daud Vicary, Domenico Del Re, Lt. Cl. Peter Garretson, Dr. Salvatore Russo, Giotto Castelli, Prof. Christine Hauskeller, and Dr. Jon Bonello, for their support, and for providing reviews for *The Space Value of Money* book.

I am grateful to the following individuals for their direct and/or indirect contributions, recently or in the past:

Prof. Dame Sandra Dawson, Dr. Mark Carney, Prof. Geoff Meeks, Prof. Arnoud De Meyer, Prof. Ha-Joon Chang, Dr. Robin Chatterjee, Dr. Jose Gabriel Palma, Prof. Tony Lawson, Prof. Geoffrey Hodgson, Dr. Rob Wallach, Prof. Peter Nolan, Prof Shailaja Fennell, Prof. Richard Barker, Dr. Rachel Armstrong, Charles Goldsmith, Erin Hallett, Sandie Campin, Ruth Newman, Jane Kemp, Jane Playdon, and Tony Manwaring.

I am grateful to the many colleagues, students, friends, and family who have contributed to the wealth and depth of my learning and experiences over the years.

While all are to be thanked, mistakes remain my own.

CONTENTS

About the Author

Armen V. Papazian is a financial economist, board director, consultant, and innovator with a track record in global finance. A former lecturer in finance, senior stock exchange executive, investment banker, and entrepreneur, Armen's career has bridged industry and academia.

He has more than 20 years of experience in sustainable finance, capital markets, and analytics. A Doctor of Financial Economics from Cambridge University, Armen combines extensive industry experience in financial institutions and markets with in-depth research into both the theoretical and practical aspects of sustainable finance.

He is the author of *The Space Value of Money*, a founder and director of the Space Value Foundation, and an active contributor to the public debate on sustainability in finance.

The Space Value of Money
Book Launch Event on the 2nd of November 2022
Keynes' Lecture Theatre—King's College Cambridge
In partnership with the Cambridge Judge Business School

ABBREVIATIONS

APF	Asset Purchase Facility
APT	Arbitrage Pricing Theory
BIM	Biodiversity Impact Metric
CAA	Climate Ambition Alliance
CAPM	Capital Asset Pricing Model
CBD	Convention on Biological Diversity
CCC	Climate Change Committee
CDE	Carbon Dioxide Equivalency
CDO	Collateralised Debt Obligations
CDP	Carbon Disclosure Project
CDSB	Climate Disclosure Standards Board
CE	Credit Easing
CGFI	UK Centre for Greening Finance and Investment
CGFI-SFI	UK Centre for Greening Finance and Investment, Spatial Finance Initiative
CISL	Cambridge Institute for Sustainability Leadership
COP26	UN's 26th Conference of the Parties
CPI	Climate Project Initiative
CSRD	Corporate Sustainability Reporting Directive
DCF	Discounted Cash Flow
DDM	Dividends Discount Model
DOJ	Department of Justice
EA	Environmental Agency
EIO-LCA	Economic Input Output—Life Cycle Assessment
ESG	Environmental, Social, and Governance
ESRS	European Sustainability Reporting Standards

ETC	Energy Transition Commission
FCA	Financial Conduct Authority
FCF	Free Cash Flows
FCFE	Free Cash Flows for Equity
FCFF	Free Cash Flows for Firm
FSB-TCFD	Financial Stability Board—Task Force on Climate-related Financial Disclosures
GDI	Green Design Institute
GEO	Geostationary Orbit
GFANZ	Glasgow Financial Alliance for Net Zero
GHG	Greenhouse Gas
GIIN	Global Impact Investing Network
GRI	Global Reporting Initiative
GSV	Gross Space Value
GTP	Global Temperature Potential
GWP	Global Warming Potential
HRC	Habitat Replacement Costs
IEA	International Energy Agency
IFRS	International Financial Reporting Standards
IPBES	Intergovernmental Science-Policy Platform on Biodiversity and Ecosystem Services
IPCC	Intergovernmental Panel on Climate Change
IRR	Internal Rate of Return
ISSB	International Sustainability Standards Board
LCA	Life Cycle Assessment
LEO	Low Earth Orbit
MEO	Medium Earth Orbit
NPV	Net Present Value
NSV	Net Space Value
NZAM	Net Zero Asset Managers Initiative
NZE	Net Zero Emissions
OECD	Organisation for Economic Co-operation and Development
ORS	Online Response System
PA	Paris Agreement
PCN	Public Capitalisation Notes
PRI	Principles of Responsible Investment
QE	Quantitative Easing
RTZ	Race to Zero
SASB	Sustainability Accounting Standards Board
SBTi	Science Based Targets initiative
SDGs	Sustainable Development Goals
SF	Sustainable Finance
SI	International System of Units

SIIT	Social Impact Investment Taskforce
SVM	Space Value of Money
TCFD	Task Force on Climate-related Financial Disclosures
TCFD-PAT	TCFD, Portfolio Alignment Team
TCRE	Transient Climate Response to Cumulative CO_2 Emissions
TNFD	Task Force on Nature-related Financial Disclosures
UNEP	United Nations Environment Programme
UNEPFI	United Nations Environment Programme Finance Initiative
UNFCCC	United Nations Framework Convention on Climate Change
UNGC	United Nations Global Compact
UNPRI	United Nations Principles of Responsible Investment
VE	Value Easing
VRF	Value Reporting Foundation
WACC	Weighted Average Cost of Capital

LIST OF FIGURES

LIST OF TABLES

CHAPTER 1

Introduction

Abstract This chapter introduces the main theme of the book which makes the case for the necessity to hardwire sustainability into financial mathematics, specifically our equations of value and return, and discusses the implications of such a transformation for money mechanics. The growing momentum in sustainable finance, while encouraging, has not yet penetrated the analytical content of finance theory and our risk and time focused equations remain intact. Furthermore, the debate seems to consider the logic and principles of money creation as exogenous to the sustainability challenge and opportunity. The book identifies the need to rethink and revise our spaceless financial value framework and equations where space, as analytical dimension and our physical context, along with our responsibility for space impact, has been abstracted away. The space value of money principle and ensuing equations are proposed as a plausible avenue through which we can achieve the necessary transformations and secure the sustainability and future expansion of human productivity across time and space.

Keywords Sustainability · Financial Mathematics · Money · Value · Risk · Time · Space · Impact

JEL Classification E00 · E58 · G00 · G30 · Q51

1

The evidence confirming human responsibility for climate change has been overwhelming (IPCC 2013, 2018, 2021, 2022). In a recent report IPCC (2023) summarises the challenge as follows:

> Human activities, principally through emissions of greenhouse gases, have unequivocally caused global warming, with global surface temperature reaching 1.1 °C above 1850–1900 in 2011–2020. Global greenhouse gas emissions have continued to increase, with unequal historical and ongoing contributions arising from unsustainable energy use, land use and land-use change, lifestyles and patterns of consumption and production across regions, between and within countries, and among individuals. (IPCC 2023, 4)

Since 2015, the targets of the Paris Agreement (UNFCCC 2015), to keep world temperature increases below *2 °C* above preindustrial levels and ideally limit the temperature increase to *1.5 °C*, have become an integral part of daily business and financial rhetoric. The operationalisation of sustainability in our daily institutional and organisational practices has become a defining theme across many fields and industries. Responding to the challenge, the finance industry and discipline have been actively developing a number of standards, frameworks, and tools for that purpose.

One of the key frameworks for the alignment of investment portfolios with our climate targets is the voluntary TCFD framework focused on the risks, opportunities, and financial impact of climate change, proposed by the Financial Stability Board—Task Force on Climate-related Financial Disclosures (TCFD 2017, 2022).[1] It is aimed at helping the industry self-regulate and adjust to the needs and requirements of the transition to a Net Zero economy (IEA 2017, 2021). Simultaneously, going beyond emissions as such, growing attention is being rightfully directed at nature and biodiversity loss with the development of the Task Force on Nature-related Financial Disclosures and the TNFD framework (IPBES 2019; TNFD 2021, 2023; CISL 2020).

In parallel, and very recently, the International Sustainability Standards Board (ISSB) released its inaugural standards (IFRS 2023a, 2023b). The Financial Stability Board (FSB) has now invited the ISSB to take

---

[1] The TCFD framework remains a central building block and its importance and relevance are unaffected by the recent developments following the publication of the IFRS S1 and S2 standards.

up TCFD's monitoring role to track progress on climate-related disclosures.[2] The IFRS standards have now been endorsed by the International Organisation of Securities Commissions (IOSCO). While both the TCFD framework and the IFRS standards are still voluntary initiatives, these developments indicate growing momentum towards a future where they will be mandatory. Indeed, on the 31st of July, the European Commission adopted the European Sustainability Reporting Standards (ESRS) to be used under the Corporate Sustainability Reporting Directive (CSRD) (EU 2023a, 2023b, 2023c). In parallel, in the UK, the Transition Plan Taskforce recently released its own framework (TPT 2023), building upon the IFRS standards and the transition planning recommendations and guidance of the Glasgow Financial Alliance for Net Zero (GFANZ 2022).

The initiatives are many, from summary buzzwords like ESG (Environmental, Social, and Governance factors), to frameworks, standards, tools, and alliances, we are now faced with a plethora of acronyms to choose from: TCFD, TNFD, ISSB, TPT, SECR, SASB, NZBA, CDP, GRI, SSE, RTZ, NZIA, NZAOA, NZAMI, PRI, etc.[3] While all these initiatives are positive developments and aim to contribute to our transition to a more sustainable world, we are still, it seems, a few steps short of effective change of trajectory.

Indeed, just as an example, a bank that supports many of the above-mentioned acronyms and their initiatives, HSBC, was recently in the news for extending "a revolving credit facility to an energy company that is tearing down a German village to expand a large coal mine, despite its promise to 'phase down' fossil fuel financing" (Martinez 2023). Similarly, on the 31st of July 2023, almost coinciding with the release of the

---

[2] The Financial Stability Board's Task Force on Climate-related Financial Disclosures recently published its latest TCFD status report in October 2023 (TCFD 2023), where they announced that "the ISSB standards represent a culmination of the Task Force's work and that the TCFD would be disbanded upon release of its 2023 status report" (TCFD 2023, ii).

[3] Task Force on Climate-related Financial Disclosures, Taskforce on Nature-related Financial Disclosures, International Sustainability Standards Board, Transition Plan Taskforce, Streamlined Energy and Carbon Reporting, Sustainability Accounting Standards Board, Net-Zero Banking Alliance, Glasgow Financial Alliance for Net Zero, Carbon Disclosure Project, Global Reporting Initiative, Sustainable Stock Exchanges initiative, Race To Zero, Net Zero Insurance Alliance, Net Zero Asset Owners Alliance, Net Zero Asset Managers Initiative, Principles of Responsible Investment.

IFRS standards, the UK Prime Minister and government announced the approval of "[h]undreds of new North Sea oil and gas licences to boost British energy independence and grow the economy" (UK Government 2023a).[4] Moreover, recent figures published by the International Monetary Fund reveal that fossil fuel subsidies have reached a record $7 trillion in 2022 (Black et al. 2023).

At COP26 in Glasgow, in November 2021, the UK's Chancellor of the Exchequer at that time made the announcement that we must rewire the global financial system for Net Zero. Indeed, to truly rewire the global financial system, to reinvent human productivity, we must understand how we ended up in our current predicament—why we have tolerated such levels of pollution and waste in our air, rivers, oceans, land, and even outer space.

> One of the main reasons for such a suboptimal outcome can be found in our financial value framework, in the principles of finance that have governed financial education and training, our markets and investments. The root cause is in the value equations of core finance theory and practice, in the equations that have been, and still are, focused on risk and time, serving one stakeholder, the risk-averse return maximising investor. (Papazian 2022, 3)

The recent momentum in sustainable finance has not yet transformed the core equations of value and return that we teach and apply in finance theory and practice, i.e., our financial mathematics of value and return. Moreover, no new equations of value seem to have been proposed that can replace our old models. Indeed, ESG integration into the investment value chain implies making adjustments to variables in our existing models (PRI 2016, 2023; PRI-CFA 2018; Papazian 2022).

---

[4] On the 20th of September 2023, in a much debated announcement, the UK Prime Minister made a number of changes to the Net Zero commitments made by the government without of course changing the broader strategic commitment to reach Net Zero by 2050 (UK Government 2023b). "Pragmatism, not ideology" is one of the key rationales given for the change and it is mainly aimed at addressing the concerns of an electorate facing a cost of living crisis.

Moreover, neither ESG integration nor climate finance nor impact investing nor responsible investing consider money mechanics a relevant subject. The sustainability effort/rhetoric, in finance industry and academia, seems to be focused on standards, scores, ratings, and frameworks that aim to operationalise sustainability at the level of investments, instruments, portfolios, and businesses, public or private. The logic of the value of money, and the equations that define its creation, allocation, and deployment are not part of the discussion. Money creation is assumed to be exogenous to the sustainability or ESG challenge/opportunity.

Given the theoretical and practical traditions in the finance discipline and industry, and recent market history in 2008 and 2023, given more than half a century of financial education built around risk and time, we cannot expect consistent, effective, and global change across industry and business unless we rethink the mathematics that underpins our financial and monetary decisions across this planet.

This rethink must begin with a reassessment of our financial value framework—a framework built around the risk and time value of money, serving the risk-averse return-maximising mortal investor, where a pollution-averse planet and an aspirational human society are considered exogenously to the models, as externalities, or, through a qualitative addendum on corporate social responsibility. Our new financial mathematics of value and return, therefore, must begin by making humanity and planet equal stakeholders of our financial models alongside the mortal risk-averse return-maximising investor.

To address the evolutionary challenges we have created for ourselves, and climate change is only one of them, we need to redefine the value of money beyond risk and time. In truth, we have an entire analytical dimension missing in our financial value framework, i.e., space. Our equations, geared towards assessing the value of cash flows vis-à-vis risk and time, omit the space impact of cash flows and assets in the value equations of those cash flows and assets. In other words, with the omission of space as an analytical dimension and our physical context, our models have also ignored space impact, and absolved investments and investors of their share of responsibility.

The remainder of the manuscript is divided into 5 main chapters:

Chapter 2 explores our current risk and time-based value framework in finance, built around two key principles of value—risk and return and time value of money. It identifies the key omissions and features of our mathematics of value and return that may explain our current predicament and reveals a missing dimension of analysis, our physical context, i.e., space.

Chapter 3 discusses the key developments in the sustainable finance field and argues that the frameworks, standards, scores, and tools of sustainability reporting fall short of penetrating core finance theory and do not transform our equations. While they shape the reporting of climate and sustainability related information, they do not interpret them, and they do not define their use and application by investors.

Chapter 4 introduces the missing dimension of analysis, i.e., space, and the missing principle of value, i.e., space value of money, and proposes a set of new equations that could be used to hardwire sustainability into financial mathematics.

Chapter 5 addresses the implications of such a transformation for money mechanics. If investors must abide by the space value of money principle and ensuing equations, then so must money creators, whether commercial or central banks. The chapter considers the challenges of our current architecture given a transformed value framework and offers alternative solutions.

Chapter 6 summarises the argument and concludes the book.

References

Black, S., Parry, I., and N. Vernon. 2023. Fossil Fuel Subsidies Surged to Record $7 Trillion. International Monetary Fund. https://www.imf.org/en/Blogs/Articles/2023/08/24/fossil-fuel-subsidies-surged-to-record-7-trillion. Accessed 4 September 2023.

CISL. 2020. Measuring Business Impacts on Nature: A Framework to Support Better Stewardship of Biodiversity in Global Supply Chains. Cambridge Institute for Sustainability Leadership, Cambridge. https://www.cisl.cam.ac.uk/system/files/documents/measuring-business-impacts-on-nature.pdf. Accessed 2 February 2022.

EU. 2023a. The Commission Adopts the European Sustainability Reporting Standards. European Commission. https://finance.ec.europa.eu/news/commission-adopts-european-sustainability-reporting-standards-2023-07-31_en. Accessed 31 July 2023.

EU. 2023b. Implementing and Delegated Acts—CSRD. European Commission. https://finance.ec.europa.eu/regulation-and-supervision/financial-services-legislation/implementing-and-delegated-acts/corporate-sustainability-reporting-directive_en. Accessed 31 July 2023.

EU. 2023c. Annex to Supplementing Directive 2013/34/EU of the European Parliament and of the Council as Regards Sustainability Reporting

Standards. European Commission. https://ec.europa.eu/finance/docs/level-2-measures/csrd-delegated-act-2023-5303-annex-1_en.pdf. Accessed 31 July 2023.

GFANZ. 2022. Financial Institution Net-zero Transition Plans: Fundamentals, Recommendations, and Guidance. Glasgow Financial Alliance for Net Zero. https://assets.bbhub.io/company/sites/63/2022/09/Recommendations-and-Guidance-on-Financial-Institution-Net-zero-Transition-Plans-November-2022.pdf. Accessed 6 February 2023.

IEA. 2017. Energy Technology Perspectives 2017: Catalysing Energy Technology Transformations. International Energy Agency. https://iea.blob.core.windows.net/assets/a6587f9f-e56c-4b1d-96e4-5a4da78f12fa/Energy_Technology_Perspectives_2017-PDF.pdf. Accessed 2 February 2021.

IEA. 2021. Net Zero by 2050: A Roadmap for the Global Energy Sector. International Energy Agency. https://iea.blob.core.windows.net/assets/deebef5d-0c34-4539-9d0c-10b13d840027/NetZeroby2050-ARoadmapfortheGlobalEnergySector_CORR.pdf. Accessed 2 February 2022.

IFRS. 2023a. IFRS S1 General Requirements for Disclosure of Sustainability-related Financial Information. IFRS. https://www.ifrs.org/issued-standards/ifrs-sustainability-standards-navigator/ifrs-s1-general-requirements/#standard. Accessed 30 July 2023.

IFRS. 2023b. IFRS S2 Climate-related Disclosures. IFRS. https://www.ifrs.org/issued-standards/ifrs-sustainability-standards-navigator/ifrs-s2-climate-related-disclosures/#standard. Accessed 30 July 2023.

IPBES. 2019. The Global Assessment Report on Biodiversity and Ecosystem Services. Intergovernmental Science-Policy Platform on Biodiversity and Ecosystem Services. https://ipbes.net/system/files/2021-06/2020%20IPBES%20GLOBAL%20REPORT%28FIRST%20PART%29_V3_SINGLE.pdf. Accessed 2 February 2021.

IPCC. 2013. Climate Change 2013: The Physical Science Basis. Summary for Policymakers. Intergovernmental Panel on Climate Change. https://www.ipcc.ch/site/assets/uploads/2018/03/WG1AR5_SummaryVolume_FINAL.pdf. Accessed 2 February 2021.

IPCC. 2018. Summary for Policymakers. In Global Warming of 1.5° C. IPCC. Available at https://www.ipcc.ch/site/assets/uploads/sites/2/2018/07/SR15_SPM_High_Res.pdf. Accessed 12 December 2020.

IPCC. 2021. Climate Change 2021: The Physical Science Basis. https://www.ipcc.ch/report/ar6/wg1/downloads/report/IPCC_AR6_WGI_SPM_final.pdf. Accessed 2 February 2022.

IPCC. 2022. Climate Change 2022: Impacts, Adaptation and Vulnerability. Summary for Policymakers. Intergovernmental Panel on Climate Change. https://report.ipcc.ch/ar6wg2/pdf/IPCC_AR6_WGII_SummaryForPolicymakers.pdf. Accessed 28 February 2022.

IPCC. 2023. Synthesis Report of the IPCC 6th Assessment Report (AR6): Summary for Policymakers. Intergovernmental Panel on Climate Change. https://report.ipcc.ch/ar6syr/pdf/IPCC_AR6_SYR_SPM.pdf. Accessed 22 March 2023.

Martinez, V. 2023. HSBC Under Fire for $340m Loan to Energy Firm Involved in Coal Mine Expansion. Investment Week, January. Accessed on 15 March 2023.

Papazian, Armen. 2022. *The Space Value of Money: Rethinking Finance Beyond Risk and Time.* New York: Palgrave Macmillan. https://doi.org/10.1057/978-1-137-59489-1

PRI. 2016. A Practical Guide to ESG Integration for Equity Investing. Principles of Responsible Investing. https://www.icgn.org/sites/default/files/2021-08/PRI_apracticalguidetoesgintegrationforequityinvesting.pdf. Accessed 30 July 2023.

PRI. 2023. ESG Integration in Listed Equity: A Technical Guide. Principles of Responsible Investing. https://www.unpri.org/download?ac=18407. Accessed 30 July 2023.

PRI-CFA. 2018. Guidance and Case Studies for ESG Integration: Equities and Fixed Income. Principles of Responsible Investment and CFA Institute. https://www.unpri.org/download?ac=5962. Accessed 30 July 2023.

TCFD. 2017. Final Report: Recommendations of the Task Force on Climate-related Financial Disclosures. https://assets.bbhub.io/company/sites/60/2020/10/FINAL-2017-TCFD-Report-11052018.pdf. Accessed 2 February 2021.

TCFD. 2022. Status Report: Task Force on Climate-related Financial Disclosures. https://assets.bbhub.io/company/sites/60/2022/10/2022-TCFD-Status-Report.pdf. Accessed 28 July 2023.

TCFD. 2023. Task Force on Climate-related Financial Disclosures: 2023 Status Report. Financial Stability Board. The Task Force on Climate-related Financial Disclosures. https://www.fsb.org/wp-content/uploads/P121023-2.pdf. Accessed 18 October 2023.

TNFD. 2021. Taskforce on Nature-related Financial Disclosures. https://tnfd.global/. Accessed 2 February 2022.

TNFD. 2023. The TNFD Nature-related Risk and Opportunity Management and Disclosure Framework. Taskforce on Nature-related Financial Disclosures. https://framework.tnfd.global/wp-content/uploads/2023/03/23-23882-TNFD_v0.4_Integrated_Framework_v7.pdf. Accessed 28 June 2023.

TPT. 2023. Disclosure Framework. Transition Plan Taskforce. https://transitiontaskforce.net/wp-content/uploads/2023/10/TPT_Disclosure-framework-2023.pdf. Accessed 18 October 2023.

UK Government. 2023a. Hundreds of New North Sea oil and Gas Licences to Boost British Energy Independence and Grow the Economy. UK Government. https://www.gov.uk/government/news/hundreds-of-new-north-sea-oil-and-gas-licences-to-boost-british-energy-independence-and-grow-the-economy-31-july-2023. Accessed on 31 July 2023.

UK Government. 2023b. PM speech on Net Zero: 20 September 2023. UK Government. https://www.gov.uk/government/speeches/pm-speech-on-net-zero-20-september-2023. Accessed 10 October 2023.

UNFCCC. 2015. Paris Agreement. United Nations Framework Convention on Climate Change. https://unfccc.int/sites/default/files/english_paris_agreement.pdf. Accessed 2 December 2020.

CHAPTER 2

The Risk and Time Value of Money

Abstract This chapter discusses our current financial value paradigm and analytical framework built around risk and time parameters, serving one stakeholder, the mortal risk-averse return-maximising investor. Our current value framework in finance theory and practice is structured around two principles of value, risk and return, and time value of money. These principles discriminate against our evolutionary investments given their internal biases towards highly risky and very distant cash flows. Furthermore, the discussion identifies a missing analytical dimension in finance, i.e., space, our physical context, and reveals space-less equations, often focused on future non-actual expected cash flows, while omitting the actual space impact it would take to achieve or expect them. Thus, sustainability in finance must surely address these omissions and make room for planet and humanity as equal stake-holders in the value and return equations that have shaped our markets and investments for the last many decades.

Keywords Sustainability · Financial mathematics · Money · Value · Risk · Time · Space · Impact

JEL Classification E00 · E58 · G00 · G30 · Q51

A. V. Papazian, *Hardwiring Sustainability into Financial Mathematics*, https://doi.org/10.1007/978-3-031-45689-3_2

Since the early beginnings and for many decades, the value framework of the finance discipline has been built to serve the mortal risk-averse return-maximising investor. The human collective and the planet have been exogenous to our models, as externalities in the economics literature,[1] and as corporate social responsibility (CSR) in the finance and management literatures. This selective focus and abstraction have defined our principles and our equations of value and return, revealing a value paradigm built around risk and time parameters (Papazian 2022).

Thomas Kuhn defines the paradigms that emerge within scientific fields as follows, they are "universally recognized scientific achievements that for a time provide model problems and solutions to a community of practitioners" (Kuhn 1962, viii). Indeed, he goes on to argue that "no natural history can be interpreted in the absence of at least some implicit body of intertwined theoretical and methodological belief that permits selection, evaluation, and criticism" (Kuhn 1962, 16).

This chapter aims to demonstrate that a survey of finance textbooks, academic and industry literature, and value and return models reveals a *risktime* analytical universe without space. Indeed, space, as an analytical dimension, and our physical context, is abstracted away from the core equations that have defined the field and its evolution throughout the last many decades.

2.1   The Risk and Time Value of Cash Flows

The risk and time focus of the discipline is explained and mirrored by the two key principles of value that have shaped finance theory, practice, education, and research. Indeed, the analytical value framework of the discipline has been built around two principles of value: (1) Risk and Return, and (2) Time Value of Money (see Table 2.1).

In the finance literature, the main stakeholder is defined as *the risk-averse investor* aiming to minimise risks and maximise returns. There is no mention of the investor's mortality. I have added the adjective 'mortal' to better describe the mindset of the main stakeholder of finance theory and mathematics.

---

[1] William Nordhaus (2018) in his Nobel Prize speech states: "Global warming is the most signifcant of all environmental externalities." (Nordhaus 2018, 441)

Table 2.1 The core principles of finance theory and practice

Stakeholder	Risk	Time
Risk-averse, return-maximising, mortal investor	**Risk and Return**: The higher the risk the higher the expected return—given the risk-averse nature of investors, higher risks imply higher expectations of reward	**Time Value of Money**: A dollar ($1) today is worth more than a dollar ($1) tomorrow—because a dollar today can earn interest/return by tomorrow and be more than a dollar by tomorrow

Source Author

This is so because risk and time are very mortal concerns. An immortal investor would be far-less concerned with time and/or risk. This is important to note because from the perspective of the human collective, which can procreate and secure its continuous existence, evolutionary continuity in space would most likely take precedence to individual risk and time concerns.

A review of finance textbooks and academic and industry literature can confirm the above observations. Brealey et al. (2020), a 13th edition core textbook in corporate finance, built on the wider academic literature, is a typical example. Similarly, in Pike et al. (2018), a 9th edition textbook on corporate finance and investment, and in Watson and Head (2016), a 7th edition on principles and practice handbook for corporate finance, we observe the same framework and principles at work.

In the professional banking and finance literature (Choudhry 2012, 2018), in investment valuation (Damodaran 2012, 2017) and company valuation (Koller et al. 2015, 2011), in project finance (Yescombe 2014), in investment banking (Rosenbaum and Pearl 2013), in property valuation and investment (Isaac and O'Leary 2013), we can see evidence of the same.

All of the above are based on the foundational academic literature where the same can be observed in our models and equations of value and return (Williams 1938; Graham and Dodd 1934; Graham 1949; Gordon and Shapiro 1956; Markowitz, 1952; Modigliani and Miller 1958, 1963; Ross 1976, 1978; Sharpe 1963, 1964; Lintner 1965; Reinganum 1981; Lakonishok and Shapiro 1986; Fama and French 1992, 1993, 1996, 2004, 2015; Black and Scholes 1973; and many others).

2.1.1   *Risk, Time, and Our Evolutionary Investments*

The risk and return and time value of money principles have defined and continue to define the analytical content and equations of finance theory and practice. In truth, these principles of value discriminate against our evolutionary investments. The biases these principles introduce are revealed through a basic profiling of the risk and time features of our evolutionary challenges/investments—the Net Zero transition is one such challenge, but it is not the only one. All our evolutionary challenges require massive investments in the present, have *very high risks*, and imply *distant returns*—features that are negatively priced based on the current principles of value that underpin the equations we use and teach in the field.

The current principles of value in finance theory leave our evolutionary investments in a blind spot. By negatively pricing distant returns and high risks, our financial value framework misprices our evolutionary investments. In fact, today, our evolutionary investments become plausible and 'affordable' only when they can be made to make sense within the preference framework of the mortal risk-averse return-maximising investor. A theoretical and practical misconception that could well explain our current predicament.

2.2   Equations Without Space, Without Context or Impact

The internal biases introduced by the current principles of value in finance identify the need to revise the framework and equations with a much broader perspective. After all, how can we even begin to address sustainability (in ESG form or other), if our core principles and equations serve an individual mortal in the chain but work against our collective purpose, human evolution.

To identify what is missing in our framework and mathematics of value and return, this section takes a closer look at our equations. To start at the very beginning, we must consider one of the most commonly used methods of valuation in finance, i.e., cash flow discounting. This, interestingly, takes us to the thirteenth century. While financial mathematics uses discounting in almost all value models across different lines of application, it does not always reveal the origin of this key valuation method.

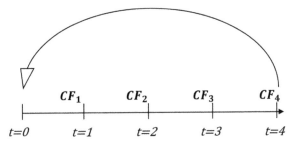

Fig. 2.1 Discounting future expected cash flows (*Source* Author)

A recent working paper by Goetzmann (2004) tracks the discounting method to Leonardo of Pisa or Fibonacci in 1202. Goetzmann finds evidence that in his Liber Abaci, *The Book of Calculation*, Fibonacci (1202) was the first to develop the present value analysis for comparing the economic value of alternative contractual or expected cash flows (Fig. 2.1).

Today, across stock, bond, project, firm, and all cash flow valuation equations taught and used in the finance discipline and industry, discounting is used to measure the time and risk value of money. The risk is introduced and 'quantified' through the use of a discount rate that represents the market rate or rate of return on an alternative investment with the same level of risk as the instrument/opportunity being assessed.

Table 2.2 lists a sample of bond, stock, asset, firm, option, and cash flow valuation equations. Naturally, this is not a comprehensive list of models and equations in finance, but it is a representative one. Indeed, many of them have defined the course of theoretical development in the field. While not all use discounting (like CAPM and other asset and option pricing models), they are all built around risk and time, using proxies for both to value the expected future cash flows and returns.

The equations in Table 2.2 reveal a financial mathematics of value and return without any contextual parameters. Our financial mathematics seems to be missing the analytical dimension of *where*. Cash flows are assessed in a *risktime* universe, abstracted from our physical context, serving the risk-averse mortal investor, without any direct mathematical reference to planet and humanity.

Table 2.2 Sample core finance equations: bonds, firms, stocks, assets, options

---

Sample bond valuation equations

$$\text{Bond Price} = \sum_{i=1}^{n} \frac{C_i}{(1+r)^n} + \frac{F}{(1+r)^n}$$

$$\text{Bond Price} = C \times \left(\frac{1 - \left(\frac{1}{(1+r)^n} \right)}{r} \right) + \frac{F}{(1+r)^n}$$

$$\text{Bond Price} = \sum_{t=1}^{n} \frac{CF_t}{(1+r)^t} + \frac{P}{(1+r)^n}$$

$$\text{Bond Price} = \sum_{t=1}^{n \times m} \frac{CF_t}{\left(1+\left(\frac{r}{m}\right)\right)^t} + \frac{P}{\left(1+\left(\frac{r}{m}\right)\right)^{n \times m}}$$

$$Bond\,Price = \left(\frac{C}{m}\right) \times \left(\frac{1 - \left(\frac{1}{(1+\frac{r}{m})^{n \times m}} \right)}{\left(\frac{r}{m}\right)} \right) + \frac{P}{(1+\frac{r}{m})^{n \times m}}$$

Sample of stock and firm valuation equations

$$P_0 = \frac{D_1}{r-g}$$

$$P_0 = \sum_{t=1}^{\infty} \frac{D_t}{(1+r)^t}$$

$$P_0 = \sum_{t=1}^{n} \frac{D_t}{(1+WACC)^t} + \frac{P_n}{(1+WACC)^n}$$

$$P_0 = \sum_{t=1}^{n} \frac{D_t}{(1+WACC)^t} + \frac{D_{n+1}}{(WACC-g).(1+WACC)^n}$$

$$\text{Firm Value} = \sum_{t=1}^{n} \frac{D_t}{(1+WACC)^t} + \frac{FCFF_{n+1}}{(WACC-g).(1+WACC)^n}$$

Sample of asset pricing models

$$R_i = R_f + \beta_i \times (R_m - R_f) \qquad \text{Beta}_i = \beta_i = \frac{\text{Covariance}_{R_i, R_m}}{\text{Variance}_{R_m}}$$

$$E(R_i) - R_f = b_1\big(E(R_M) - R_f\big) + s_i\,E(SMB) + h_i\,E(HML)$$

Modigliani Miller corporate value and capital structure model

$$V_j = (S_j + D_j) = \frac{\overline{X_j}}{\rho_k}$$

$$i_j = \rho_k + (\rho_k - r)\frac{D_j}{S_j}$$

Black and Scholes option pricing model

$$C = SN(d) - Le^{-rt}N\big(d - \sigma\sqrt{t}\big)$$

(continued)

Table 2.2 (continued)

$$d = \frac{\ln \frac{S}{L} + \left(r + \frac{\sigma^2}{2}\right)t}{\sigma \sqrt{t}}$$

Net Present Value cash flow valuation model

$$NPV = \sum_{t=0}^{T} \frac{CF_t}{(1+r)^t} \qquad NPV = CF_0 + \sum_{t=1}^{T} \frac{CF_t}{(1+r)^t}$$

$$\text{Net Present Value} = -II + \sum_{t=1}^{n} \frac{CF_t}{(1+r)^t}$$

See Brealey et al. (2020), Pike et al. (2018), Watson and Head (2016), Fama and French (1996, 2004, 2015), Gordon (1959), Gordon and Gordon (1997), Gordon and Shapiro (1956), Modigliani and Miller (1958), Ross (1976), Roll and Ross (1980), Sharpe (1964), Lintner (1965), Merton (1973), Black and Scholes (1973), Nobel Prize (1997); and others. See Papazian (2022) for a detailed discussion of the above, and the absence of space and space impact.

Source Compiled by Author

In truth, our financial value framework is missing the analytical dimension of *Space*—our physical context stretching from subatomic to interstellar space and every layer in between and beyond. By and through this omission, our equations have also abstracted away our responsibility for space impact.[2]

This is the theoretical and mathematical junction where we have ignored our context and our responsibility for impact, causing unprecedented environmental degradation, a climate crisis, and a host of socioeconomic challenges. It is the level and degree of this misconception that also explains why adjusting variables with ESG considerations will prove to be an ineffective strategy and a poor conceptualisation of sustainability in finance theory and practice. I discuss and explore this argument in detail in Chapter 3.

In parallel to the equations in Table 2.2, the risk and time focus of the discipline is also revealed through the vast literature on risk (Haynes 1895; Knight 1921), portfolio optimisation (Markowitz 1952; Rom and

[2] In my unpublished doctoral dissertation (Papazian 2004), An Endoscopy on Stock Market Winners and Losers, focused on London Stock Exchange listed stocks included in the FT500 index between 1988 and 1992, I applied clinical methodology (Jensen et al. 1989) to explore and scrutinise the firms included in the portfolios in real time and space between 1988 and 1996.

Ferguson 1993), and stock market predictability—discussing market efficiency, random walks, and overreaction in the context of risk-adjusted returns (see also Fama 1970; Malkiel 1973; De Bondt and Thaler 1985; Dissanaike 1994, 1997; Harvey et al. 2016; Xi et al. 2022; and others). A similar omission can be observed in the company valuation literature (see Koller et al. 2015, 2011).

Koller et al. (2015) in *Valuation: Measuring and Managing the Value of Companies*, 6th edition, discussing the 'Fundamental Principles of Value Creation' state and summarise how this discussion applies to companies:

> Companies create value for their owners by investing cash now to generate more cash in the future. The amount of value they create is the difference between cash inflows and the cost of the investments made, adjusted to reflect the fact that tomorrow's cash flows are worth less than today's because of the time value of money and the riskiness of future cash flows. (Koller et al. 2015, 17)

The equations listed in Table 2.2 reflect the same principles of value, i.e., time value of money and risk and return, and omit space as an analytical dimension. Thus, they do not include the space impact of cash flows as an integral part of the equations of value/return.

While a detailed discussion of all of the equations listed in Table 2.2 is outside the scope of this book (see Papazian 2022), I expand on two core equations in the next sections, the Net Present Value model and the Capital Asset Pricing Model.

2.3   Discounting the Non-actual, Omitting the Actual: NPV

Our current financial value framework built around risk and time, serving the mortal risk-averse return-maximising investor, has inbuilt biases against our evolutionary investments, and has an entire dimension of context, i.e., space, missing from the equations of value and return it teaches and applies in theory and practice.

Interestingly, our financial mathematics of value and return through our models and equations, reveals yet another key architectural shortcoming—a bias towards the non-actual figures, i.e., cash flows in the future. While this in itself could be considered a harmless focus, combined with the omission of space and responsibility from our models, it can further explain our current predicament.

I discuss the Net Present Value equation here as it epitomises a risk and time-based financial mathematics. The discussion can be extended to almost all discounting models used in the field. The Net Present Value (NPV) equation, Eq. 2.1, reveals that the mathematical focus is on the imaginary future expected cash flows rather than the actual space impact of investments—which are treated only with a '−' sign to denote an outflow.[3,4]

$$\textbf{Net Present Value} = \boxed{-\ II} + \overbrace{\sum_{t=1}^{n} \frac{CF_t}{(1+r)^t}}$$

(2.1)

Actual   Non-Actual

n = Time Horizon

[3] The NPV equation is sometimes written in the below formats, where the first cash flow CF_0 (II) is included in the right-side term as the first cash flow at $t = 0$, or excluded but without the negative sign as the negative sign of the first cash flow CF_0 is assumed:

$$NPV = \sum_{t=0}^{T} \frac{CF_t}{(1+r)^t} \qquad NPV = CF_0 + \sum_{t=1}^{T} \frac{CF_t}{(1+r)^t}$$

[4] The Net Present Value equation (NPV) is one of the most commonly used equations. Indeed, Graham and Harvey (2002) reveal that Net Present Value (NPV) is one of the most frequently used capital budgeting techniques by Chief Financial Officers (CFOs), along with the internal rate of return (IRR), which is the discount rate that equalises NPV to zero.

t = Moving time
r = Discount Rate
II = Initial Investment
CF_t = Future Expected Cash Flows

> The omission of space happens when we ascribe an abstract negative sign to the initial investment, disregarding it entirely as a real process with a space impact. Indeed, the assessment of impact can be achieved by digging deeper into the many aspects of the investment and its deployment. In other words, the omission of space and space impact occurs when our treatment of the initial investment is defined and limited to ascribing a negative sign to denote an outflow for the risk-averse investor without further analysis or consideration of its utilisation and impact on the environment, on society, on space. (Papazian 2022, 24)

There are two parts to the NPV equation: the first part, the actual part, which is what we would be investing to be able to expect the future expected cash flows, the initial investment (II), and the second part, the non-actual part, the expected future cash flows (CF_t). Future expected cash flows are non-actual or imaginary because they have not happened yet. They may happen as expected or agreed, or they may not. If these cash flows were guaranteed, there would be no need to discount them into the present to account for their riskiness over time. Naturally, applying a discount rate to the future expected cash flows does not make the cash flows any less non-actual, or more real (Papazian 2022, 24).

The certain element in the NPV equation is the initial investment (II), and it is treated with a negative sign. This indicates an outflow for the investor and denotes the abstraction of context and space. Meanwhile, the non-actual part is mathematically treated for time and risk. This reveals a mathematical attention entirely focused on the expected cash flows in the future, while the investment, that which is most certain in the present, is treated simply with a minus sign. The impact of the initial investment is abstracted away along with the dimension of context and space. Our models are unconcerned with the impact it takes to achieve the future expected cash flows.

2.4 Risk as Time-Based Performance: CAPM

The Capital Asset Pricing Model (CAPM) (Sharpe 1964; Lintner 1965) is one of the core asset pricing models used and applied in finance. Its role in shaping asset pricing theory and practice is beyond any doubt. It is the subject matter of a vast literature, with numerous critiques, many variations, and a body of empirical evidence that support and refute the model based on past empirical correlations.

Fama and French (2004) describe the relevance of the model as follows:

> The capital asset pricing model (CAPM) of William Sharpe (1964) and John Lintner (1965) marks the birth of asset pricing theory (resulting in a Nobel Prize for Sharpe in 1990). Four decades later, the CAPM is still widely used in applications, such as estimating the cost of capital for firms and evaluating the performance of managed portfolios. It is the centrepiece of MBA investment courses. Indeed, it is often the only asset pricing model taught in these courses. (Fama and French 2004, 25)

The CAPM constructs a relationship between expected returns on a security and its riskiness. It is built on the risk and return principle, and although there is no discounting involved in the model, time plays a central role, revealing once again the risk and time focus of our analytical framework and resulting equations.

The model proposes that the return on a security i, R_i, is equal to the risk-free rate, R_f, which is conceptually the return on an investment with zero risk and is usually considered to be the return on a government bond or T-Bill, plus a reward for the risk being taken by being in that asset or security. The reward is equal to the market premium, $R_m - R_f$, multiplied by a systematic risk proxy of the security i, Beta, β_i.

$$R_i = R_f + \beta_i \times \left(R_m - R_f\right) \tag{2.2}$$

R_i = Return on security i
R_f = Risk Free Rate
β_i = Risk proxy
R_m = Return on market

The elements of the model reveal that there is no reference to the impact of the asset on our planet, on humanity, on space. There is no context to the expectations of return, which are assumed to be dependent on risk, Beta, which is a relative volatility measure vis-à-vis the market, represented through a market index.

$$\text{Beta}_i = \beta_i = \frac{\text{Covariance}_{R_i, R_m}}{\text{Variance}_{R_m}} \tag{2.3}$$

R_i is the return on security i, R_m is the return on the market, and the relationship is abstracted away from the space impact of the corporation or asset, and even further away is the space impact of the corporations included in the market index. In other words, the return on a security or asset is theoretically conceived to be linked to its riskiness, which is itself built on the time-based performance of the returns, R_i and R_m, and their covariance.

While there are many adjustments and variations to the CAPM, such as the ICAPM and CCAPM,[5] and many introduce behavioural considerations, tastes, and sustainability-related factors (Fama and French 2007; Zerbib 2019; Pedersen et al. 2021; Capelli et al. 2023), the principal logic being tested remains the same, a risk-return relationship with an adjusted explanation or measurements of risk and premiums.

Moreover, the dependence on the time-based performance of returns as a key proxy for the risk-return relationship remains central. There is no role for the actual space impact of the securities, and the conceptualisation and measurement of risk (Beta), does not take any space-based considerations into account.

2.5   Conclusion

Given the omission of context from our financial mathematics of value and return, given the omission of space from our equations, given the biases against our evolutionary investments, given the focus on the mortal risk-averse return-maximising investor as only stakeholder, given the mathematical attention of our models on the future non-actual expected

---

[5] Intertemporal Capital Asset Pricing Model and Consumption-based Capital Asset Pricing Model.

cash flows and the abstraction of actual space impact, can we be surprised with the current state of the world?

The carbon in our air, the plastic in our oceans, the sewage in our rivers and beaches, the chemical waste in landfills, the radioactive waste under the ocean bed, and the debris in orbit, etc., they are all evidence of the necessity to reinvent human productivity—a transformation that requires a radical rethink of our financial value paradigm and equations of value and return.

Indeed, our mathematics of value and return in finance theory and practice have continuously abstracted away our physical context, and the two key stakeholders in it, i.e., humanity and planet. By doing so, they have also conveniently omitted the responsibility for space impact from our models and thus value and return equations.

To effectively change trajectory and ensure the sustainability and continuous expansion of human productivity, we must, surely, recognise the necessity to transform a spaceless value framework where responsibility is exogenous.

It is very hard to visualise a sustainable world when the equations that underpin billions of financial and monetary decisions across the planet are still based on risk and time alone, without context parameters and without consideration of impact.

REFERENCES

Black, F., and M. Scholes. 1973. The Pricing of Options and Corporate Liabilties. *The Journal of Political Economy* 81: 637–654. https://www.jstor.org/stable/1831029. Accessed 02 February 2021.

Brealey, A.R., C.S. Myers, and F. Allen. 2020. *Principles of Corporate Finance*, 13th ed. New York: McGraw Hill.

Capelli, P., Ielasi, F., and A. Russo. 2023. Integrating ESG risks into value-at-risk, *Finance Research Letters*, 55 Part A, 103875, ISSN 1544–6123, https://doi.org/10.1016/j.frl.2023.103875. Accessed 3 August 2023.

Choudhry, M. 2012. *The Principles of Banking*. Singapore: Wiley.

Choudhry, M. 2018. *Past, Present, and Future Principles of Banking and Finance*. Singapore: Wiley.

Damodaran, A. 2012. *Investment Valuation*, 3rd ed. Hoboken, NJ: Wiley.

Damodaran, A. 2017. *Damodaran on Valuation*, 2nd ed. Hoboken, NJ: Wiley.

De Bondt, W.F.M., and R. Thaler. 1985. Does the Stock Market Overreact? *The Journal of Finance* 40 (3): 793–805. https://doi.org/10.2307/2327804. Accessed 02 February 2021.

Dissanaike, G. 1994. On the Computation of Returns in Tests of the Stock Market Overreaction Hypothesis. *Journal of Banking & Finance* 18 (6): 1083–1094. https://doi.org/10.1016/0378-4266(94)00061-1. Accessed 02 February 2022.

Dissanaike, G. 1997. Do Stock Market Investors Overreact? *Journal of Business Finance and Accounting* 24 (1): 27–50. https://doi.org/10.1111/1468-5957.00093. Accessed 02 February 2021.

Fama, E.F. 1970. Efficient Capital Markets: A Review of Theory and Empirical Work. *The Journal of Finance* 25: 383–417.

Fama, E.F., and K.R. French. 1992. The Cross-Section of Expected Stock Returns. *The Journal of Finance* 47: 427–465. https://doi.org/10.2307/2329112. Accessed 02 February 2021.

Fama, E.F., and K.R. French. 1993. Common Risk Factors in the Returns on Stocks and Bonds. *The Journal of Financial Economics* 33: 3–56. https://doi.org/10.1016/0304-405X(93)90023-5. Accessed 02 February 2021.

Fama, E.F., and K.R. French. 1996. Multifactor Explanations of Asset Pricing Anomalies. *The Journal of Finance* 51: 55–84. https://doi.org/10.1111/j.1540-6261.1996.tb05202.x. Accessed 02 February 2021.

Fama, E.F., and K.R. French. 2004. The Capital Asset Pricing Model: Theory and Evidence. *Journal of Economic Perspectives* 18: 25–46. https://www.aeaweb.org/articles?id=10.1257/0895330042162430. Accessed 02 February 2021.

Fama, E.F., and K.R. French. 2007. Disagreement, tastes, and asset prices, *Journal of Financial Economics*, 83 (3): 667–689, ISSN 0304-405X, https://doi.org/10.1016/j.jfineco.2006.01.003. Accessed 20 July 2023.

Fama, E.F., and K.R. French. 2015. A Five-Factor Asset Pricing Model. *Journal of Financial Economics* 116: 1–22. https://doi.org/10.1016/j.jfineco.2014.10.010. Accessed 02 February 2021.

Fibonacci, Leonardo of Pisa. 1202. *Liber Abaci*. Translated by Sigler, L.E. 2002. *Fibonacci's Liber Abaci*. Springer-Verlag.

Goetzmann, W.N. 2004. Fibonacci and the Financial Revolution. National Bureau of Economic Research. Working Paper 10352. http://www.nber.org/papers/w10352. Accessed 02 February 2022.

Gordon, M.J. 1959. Dividends, Earnings, and Stock Prices. *The Review of Economics and Statistics* 41: 99–105. https://doi.org/10.2307/1927792. Accessed 02 February 2021.

Gordon, J.R., and M.J. Gordon. 1997. The Finite Horizon Expected Return Model. *Financial Analysts Journal* 53: 52–61. https://doi.org/10.2469/faj.v53.n3.2084. Accessed 02 February 2021.

Gordon, M.J., and E. Shapiro. 1956. Capital Equipment Analysis: The Required Rate of Profit. *Management Science* 3: 102–110. https://www.jstor.org/stable/2627177. Accessed 02 February 2021.

Graham, B. 1949. *The Intelligent Investor*. New York: Harper & Brothers.

Graham, B., and D. Dodd. 1934. *Security Analysis*. New York: McGraw Hill.

Graham, J., and C. Harvey. 2002. How CFOs Make Capital Budgeting and Capital Structure Decisions. *Journal of Applied Corporate Finance* 15: 8–23. https://doi.org/10.1111/j.1745-6622.2002.tb00337.x. Accessed 02 February 2021.

Harvey, R., Y. Liu, and H. Zhu. 2016. ... and the Cross-Section of Expected Returns. *The Review of Financial Studies* 29: 5–68. https://doi.org/10.1093/rfs/hhv059. Accessed 02 February 2021.

Haynes, J. 1895. Risk as an Economic Factor. *The Quarterly Journal of Economics* 9 (4): 409–449. https://doi.org/10.2307/1886012. Accessed 02 February 2021.

Isaac, D., and J. O'Leary. 2013. *Property Valuation Techniques*, 3rd ed. London: Palgrave Macmillan.

Jensen, M.C., Fama, E., Long, J., Ruback, R., Schwert, G.W., and J.B. Warner, 1989. Editorial: clinical papers and their role in the development of financial economics. *Journal of Financial Economics* 24: 3–6. https://www.sciencedirect.com/science/article/abs/pii/0304405X8990069X?via%3Dihub. Accessed 02 February 2021.

Knight, F.H. 1921. *Risk, Uncertainty and Profit*. New York: Houghton Mifflin Company.

Koller, T., R. Dobbs, B. Huyett, McKinsey and Company. 2011. *Value: The Four Cornerstones of Corporate Finance*, 6th ed. Hoboken, NJ: Wiley.

Koller, T., M. Goedhart, D. Wessels, McKinsey and Company. 2015. *Valuation: Measuring and Managing the Value of Companies*, 6th ed. Hoboken, NJ: Wiley.

Kuhn, T. 1962. *The Structure of Scientific Revolutions*. Chicago: University of Chicago Press.

Lakonishok, J., and A.C. Shapiro. 1986. Systematic Risk, Total Risk and Size as Determinants of Stock Market Returns. *Journal of Banking & Finance* 10: 115–132. https://doi.org/10.1016/0378-4266(86)90023-3. Accessed 02 February 2021.

Lintner, J. 1965. The Valuation of Risk Assets and the Selection of Risky Investments in Stock Portfolios and Capital Budgets. *The Review of Economics and Statistics* 47: 13–37. https://doi.org/10.2307/1924119. Accessed 02 February 2021.

Malkiel, B.G. 1973. *A Random Walk Down Wall Street*. New York: W. W. Norton.

Markowitz, H. 1952. Portfolio Selection. *The Journal of Finance* 7: 77–91. https://doi.org/10.2307/2975974. Accessed 02 February 2021.

Merton, R. 1973. An Intertemporal Capital Asset Pricing Model. *Econometrica* 41: 867–887. https://doi.org/10.2307/1913811. Accessed 02 February 2021.

Modigliani, F., and M.H. Miller. 1958. The Cost of Capital, Corporation Finance and the Theory of Investment. *The American Economic Review* 48: 261–297. https://www.jstor.org/stable/1809766. Accessed 02 February 2021.

Modigliani, F., and M.H. Miller. 1963. Corporate Income Taxes and the Cost of Capital: A Correction. *The American Economic Review* 53: 433–443. https://www.jstor.org/stable/1809167. Accessed 02 February 2021.

Nobel Prize. 1997. For a New Method to Determine the Value of Derivatives. The Nobel Prize. Press Release. https://www.nobelprize.org/prizes/economic-sciences/1997/press-release/. Accessed 02 February 2022.

Nordhaus, W.D. 2018. Climate Change: The Ultimate Challenge for Economics. Nobel Prize Lecture. https://www.nobelprize.org/uploads/2018/10/nordhaus-lecture.pdf. Accessed 02 February 2022.

Papazian, A.V. 2004. *An Endoscopy on Stock Market Winners and Losers. Unpublished PhD Dissertation.* Cambridge University Judge Business School. UK: Cambridge.

Papazian, Armen. 2022. *The Space Value of Money: Rethinking Finance Beyond Risk and Time.* New York: Palgrave Macmillan. https://doi.org/10.1057/978-1-137-59489-1.

Pedersen, L.H., Fitzgibbons, S., and L. Pomorski, 2021. Responsible investing: The ESG-efficient frontier, Journal of Financial Economics, 142 (2): 572–597, ISSN 0304-405X, https://doi.org/10.1016/j.jfineco.2020.11.001. Accessed 6 September 2023.

Pike, R., B. Neale, S. Akbar, and P. Linsley. 2018. *Corporate Finance and Investment*, 9th ed. London: Pearson.

Reinganum, M.R. 1981. Misspecification of Capital Asset Pricing: Empirical Anomalies Based on Earnings' Yields and Market Values. *Journal of Financial Economics* 9: 19–46. https://doi.org/10.1016/0304-405X(81)90019-2. Accessed 02 February 2021.

Roll, R., and S.A. Ross. 1980. An Empirical Investigation of the Arbitrage Pricing Theory. *The Journal of Finance* 35: 1073–1103.

Rom, B.M., and K. Ferguson. 1993. Post-modern Portfolio Theory Comes of Age. *Journal of Investing* 3: 11–17. https://doi.org/10.3905/joi.2.4.27. Accessed 02 February 2021.

Rosenbaum, J., and J. Pearl. 2013. *Investment Banking*. Hoboken, NJ: Wiley.

Ross, S.A. 1976. The Arbitrage Theory of Capital Asset Pricing. *Journal of Economic Theory* 13: 341–360. https://doi.org/10.1016/0022-0531(76)90046-6. Accessed 02 February 2021.

Ross, S.A. 1978. The Current Status of the Capital Asset Pricing Model (CAPM). *Journal of Finance* 33: 885–890. https://doi.org/10.2307/2326486. Accessed 02 February 2021.

Sharpe, W.F. 1963. A Simplified Model for Portfolio Analysis. *Management Science* 9: 277–293. https://www.jstor.org/stable/2627407. Accessed 02

February 2021.

Sharpe, W.F. 1964. Capital Asset Prices: A Theory of Market Equilibrium Under Conditions of Risk. *Journal of Finance* 19: 425–442. https://doi.org/10.1111/j.1540-6261.1964.tb02865.x. Accessed 02 February 2021.

Watson, D., and A. Head. 2016. *Corporate Finance: Principles and Practice*, 7th ed. London: Pearson.

Williams, J.B. 1938. *The Theory of Investment Value*. Cambridge: Harvard University Press.

Xi, D., L. Yan, D.E. Rapach, and G. Zhou. 2022. Anomalies and the Expected Market Return. *The Journal of Finance* 77: 639–681. https://doi.org/10.1111/jofi.13099. Accessed 20 February 2021.

Yescombe, E.R. 2014. *Principles of Project Finance*, 2nd ed. Oxford: Academic Press.

Zerbib, O.D. 2019. The effect of pro-environmental preferences on bond prices: Evidence from green bonds, *Journal of Banking & Finance*, Volume 98, 2019, Pages 39-60, ISSN 0378-4266, https://doi.org/10.1016/j.jbankfin.2018.10.012. Accessed 10 march 2023.

CHAPTER 3

Sustainability in Finance: Frameworks, Standards, and Scores

Abstract This chapter discusses a number of key developments in the sustainable finance field and argues that the frameworks and standards of reporting climate and sustainability related information fall short of penetrating finance theory and do not transform our core equations in the field. The standards and frameworks of sustainability, built around ESG factors, do not elaborate nor provide the broader transformations through which the reported information must be used/interpreted/applied by investors. In other words, our spaceless equations remain intact and ESG integration is tantamount to adjusting variables in our existing models. Moreover, the chapter reveals that across the many strands of the sustainable finance field today, the logic and principles that govern money mechanics, i.e., money creation, are left out of the debate and assumed to be exogenous—as if our monetary architecture is immaterial to the sustainability challenge/opportunity.

Keywords Sustainability · Financial mathematics · Money · Value · Risk · Time · Space · Impact · ESG

JEL Classification E00 · E58 · G00 · G30 · Q51

A. V. Papazian, *Hardwiring Sustainability into Financial Mathematics*, https://doi.org/10.1007/978-3-031-45689-3_3

This chapter discusses key developments in the sustainable finance field and argues that the frameworks and standards of reporting climate and sustainability related information fall short of penetrating core finance theory and do not transform our equations in the field. Indeed, their purpose is entirely different, and it is focused on structuring the information flow made available to the market.

While useful and critical for our sustainability journey, the standards and frameworks of sustainability, built around ESG factors, do not elaborate on nor provide the broader transformations through which the reported information must be used/interpreted/applied by investors. In other words, our equations of value and return discussed in Chapter 2 remain intact, i.e., our spaceless analytical framework continues to define the theory, practice, and education of finance.

In the absence of any theoretical and mathematical change to the principles and equations of value and return, ESG ratings and scores have become a popular proxy for sustainability. However, as the discussion will show, these scores are often in disagreement, they do not actually measure impact, often correlate with high negative environmental impact like emissions and pollution, and the evidence of any relationship with positive price performance is at best mixed. In truth, ESG scores have been a convenient, albeit ineffective, way to consider sustainability *without* changing our financial value framework and associated value and return equations.

Moreover, the chapter reveals that across the many strands of the sustainable finance field today, whether ESG integration, impact investing, or climate finance, we observe another critical omission. The logic of money creation and the principles that govern the creation, allocation, and deployment of money are treated as exogenous to the sustainability challenge/opportunity—as if our monetary architecture is immaterial to the challenges we face on a planetary level.

3.1   Overview

Since the Paris Agreement and the Sustainable Development Goals (UNFCCC 2015; UN 2016), the Net Zero and Race to Zero initiatives (IEA 2017, 2021; UNFCCC 2020), and the establishment of human responsibility for climate change and worldwide biodiversity loss and environmental degradation (IPCC 1988, 2013, 2018a, b, 2021, 2022, 2023; MEA 2003; IPBES 2019; CBD 2021; Dasgupta 2021), the adoption and operationalisation of sustainability in the finance industry has become a global priority. The raging climate crisis has continuously validated these efforts, and the disastrous consequences of the many widespread wildfires, heatwaves, floods, and extreme hailstorms have justified their urgency.

The conceptualisation and operationalisation of sustainability in finance has been mainly focused on developing a variety of frameworks and standards that aim to support such an objective. We have witnessed the growth and development of sustainability standards (IFRS 2021, 2023a, b; EU 2023a, b, c; IIRC 2013, 2021; SASB 2020a, b, c, 2021; GRI 2021, 2022; EFRAG 2021; SEEA 2014; WRI and WBCSD 2004, 2011) and frameworks for climate and nature-related financial disclosures that aim to support the alignment of the sector with the sustainability targets (TCFD 2017, 2021a, b; TNFD 2021a, b, c, 2023a, b).

Similarly, we have also witnessed the development of mitigation pathways, transition plans, science-based targets, ESG ratings and factor integration, portfolio alignment frameworks and metrics like temperature scores, and a host of solutions aimed at helping and implementing this transition in business, industry, and finance (TPT 2023; GFANZ 2022; SEC 2022; CCC 2019; HM Government 2019; CISL 2020, 2019, a, b; EU 2018, 2020a, b, 2022; EU-TEG 2019; CDP 2020; CDP-WWF 2020; ILB et al. 2020; SEBTi 2021, 2022; TCFD-PAT 2020, 2021; PRI 2016, 2020, 2023; MSCI 2018; S&P Global 2022a; UBS-RI 2022). Moreover, extensive effort is being invested to quantify the amount of new funding needed for the transition (HSBC-BCG 2021; CPI 2021; McKinsey GI 2022).

In parallel, banks, asset managers, insurance firms, and pension funds have initiated their own drive to support the transition through the Glasgow Financial Alliance for Net Zero (UNFCCC 2021) and other industry-focused initiatives (GFANZ 2022). GFANZ is an alliance with more than 550 members in more than 50 jurisdictions now committed to support this Net Zero target.

Amongst the many approaches and measures we observe in this evolving market, the most discussed strategy is the *alignment of companies and investment portfolios* with standards and frameworks that aim at facilitating the reporting of climate and sustainability related material information.

The pathways, targets, standards, frameworks, metrics, tools, and funding mechanisms of the transition to a sustainable Net Zero world economy are work in progress, and they are being shaped through an intense global debate affected by various local, domestic, and regional priorities.

3.2   Sustainability Reporting Standards: IFRS and ESRS

The effort to develop voluntary sustainability standards has been a work in progress since the 1980s and 1990s. One of the earliest structured propositions was made through the Global Reporting Initiative (1997), followed by many others including the Climate Disclosure Standards Board (2007), the Value Reporting Foundation (2010), the International Integrated Reporting Council (2010), and the Sustainability Accounting Standards Board (2011).

Following industry debate and academic discussions (Barker and Eccles 2018), in November 2021, during COP26, the International Financial Reporting Standards Foundation (IFRS 2021) announced that it has joined forces with the Climate Disclosure Standards Board (CDSB) and the Value Reporting Foundation (VRF—including the Integrated Reporting Framework and the Sustainability Accounting Standards Board—SASB).

In this section, I discuss the recently released IFRS S1 and S2 standards, and the recently adopted European Sustainability Reporting Standards (ESRS), to reveal their focus and purpose, and clarify that these reporting standards, however important and necessary, do not transform our core principles and equations of finance—that is not their purpose.

3.2.1   IFRS Standards S1 and S2

The International Financial Reporting Standards Foundation initiated the creation of the International Sustainability Standards Board (ISSB) in November 2021, and the announcement was made during COP26 in Glasgow. Its purpose was, and still is, to create the world's first international sustainability standards which aimed "to develop—in the public interest—a comprehensive global baseline of high-quality sustainability disclosure standards to meet investors' information needs" (IFRS 2021).

In June 2023, IFRS published its first sustainability standards. The first key paragraphs of the published ISSB sustainability standards, S1 and S2, define the key objectives:

> The objective of IFRS S1 General Requirements for Disclosure of Sustainability-related Financial Information is to require an entity to disclose information about its sustainability-related risks and opportunities that is useful to primary users of general purpose financial reports in making decisions relating to providing resources to the entity. (IFRS 2023a, 6)

> The objective of IFRS S2 Climate-related Disclosures is to require an entity to disclose information about its climate-related risks and opportunities that is useful to primary users of general purpose financial reports in making decisions relating to providing resources to the entity. (IFRS 2023b, 5)

The fact that IFRS is addressing climate and sustainability in two separate standard publications reveals the growing depth and breadth of thinking. Indeed, climate change and carbon emissions are not the only elements or aspects relevant to our sustainability drive. There are many challenges to address including environmental degradation and biodiversity loss.

The scope of the published standards is clearly defined (IFRS 2023a, b): "[s]*ustainability-related risks and opportunities that could not reasonably be expected to affect an entity's prospects are outside the scope of this Standard.*" In other words, the focus is on the disclosure of useful and material information on the entities' risk/opportunities in relation to sustainability and climate that affect their prospects and are thus relevant to the decisions of providing resources to the said entities.

While the purpose of the IFRS S1 and S2 standards is to structure and define the flow of material information relevant to '*making decisions relating to providing resources to the entity,*' these standards do not provide nor expressly require a new and transformed approach to how investors or resource providers use the reported information.

The utilisation and application of the information reported by companies, whose materiality is decided by the companies or entities themselves (IFRS 2023a, 28), remains in the realm of finance and the many financial value and return principles and equations that have guided markets and investments over the last many decades. IFRS (2023a) states this clearly:

The decisions of primary users relate to providing resources to the entity and involve decisions about:

(a) buying, selling, or holding equity and debt instruments;
(b) providing or selling loans and other forms of credit; or
(c) exercising rights to vote on, or otherwise influence, the entity's management's actions that affect the use of the entity's economic resources.

The decisions described in paragraph B14 [above] depend on *primary users' expectations about returns, for example, dividends, principal and interest payments or market price increases.*[1] *Those expectations depend on primary users' assessment of the amount, timing and uncertainty of future net cash inflows to the entity* and on their assessment of stewardship of the entity's economic resources by the entity's management and its governing body(s) or individual(s). (IFRS 2023a, 28)

It is clear from the above that risk, our existing models, and their variables discussed in Chapter 2 are still centre stage. The mentioned elements that could define the funding decisions of the primary users are still the same, expectations of returns, dividends, principal and interest payments, price increases, and riskiness.

It is beyond any doubt that sustainability accounting standards are important, and necessary. However, in their very nature, they do not go far enough to change the content of core finance principles and models—they do not change the value and return equations being taught and applied across finance theory and practice—the framework and models that define how any and all information is used and applied by investors, or resource providers.

3.2.2 ESRS: European Sustainability Reporting Standards

The European Sustainability Reporting Standards (ESRS) initiated by the European Financial Reporting Advisory Group (EFRAG 2021) aimed at the development of robust sustainability standards for the European Union. The Corporate Sustainability Reporting Directive (CSRD) of the European Commission (EU 2023a, b) entered into force on the 5th of January 2023, it enhanced the sustainability reporting requirements of around 50,000 large companies and listed SMEs, and required that they

---

[1] Emphasis added.

report according to the European Sustainability Reporting Standards (EU 2023c).

> The new rules will ensure that investors and other stakeholders have access to the information they need to assess investment risks arising from climate change and other sustainability issues. They will also create a culture of transparency about the impact of companies on people and the environment. (EU 2023a)

While it is a truly inspiring development to witness such reporting requirements become mandatory, we observe that the focus is on assessing *investment risks*. The next clearly mentioned objective, termed much more loosely, is to create '*a culture of transparency about the impact of companies on people and the environment.*' This reflects the double materiality concept first officially introduced by the European Commission (EU 2019; Adams et al. 2021).

The ESRS, as announced by the European Commission, has been developed in discussions with the International Sustainability Standards Board and Global Reporting Initiative to ensure alignment between EU and global standards.

While an important milestone in sustainability reporting, like the IFRS standards, the ESRS offers standards for the reporting of sustainability related material information, it does not define or frame their use and application, nor transform our core finance principles and equations.

3.3   FRAMEWORKS AND TOOLS: TCFD, TNFD, AND ITR

The FSB-TCFD (Financial Stability Board-Task Force on Climate Related Financial Disclosures) voluntary framework is primarily focused on climate change and its purpose is "to develop voluntary, consistent climate-related financial disclosures that would be useful to investors, lenders, and insurance underwriters in understanding material risks.... its members were selected by the Financial Stability Board and come from various organisations, including large banks, insurance companies, asset managers, pension funds, large non-financial companies, accounting and consulting firms, and credit rating agencies" (TCFD 2017, iii).[2]

The TCFD has become the main finance industry framework[3] for climate-related disclosures, enjoying the support of more than 3,800 organisations in 99 countries, including 1,500 financial institutions, responsible for assets of $217 trillion (TCFD 2022). While a positive momentum is observed in the support garnered by the framework, it should be noted that given recent figures published by MSCI, less than 40% of MSCI ACWI Investable Market Index constituents reported Scope 1 and 2 emissions, and less than 25% reported Scope 3 GHG emissions (Bokern 2022).

The TCFD report that established the framework explained the rationale as follows:

> Those organizations in early stages of *evaluating the impact of climate change on their businesses and strategies* can begin by disclosing climate-related issues as they relate to governance, strategy, and risk management practices. The Task Force recognizes the challenges associated with measuring the impact of climate change but believes that by moving climate-related issues into mainstream annual financial filings, practices and techniques will evolve more rapidly. (TCFD 2017, v)[4]

The above reveals a mindset that echoes what we observed with the standards in the previous sections; the central focus is on measuring the impact of climate change on the businesses and the financial system. This is why the framework is identified as one of Climate-Related Risks,

---

[2] This discussion is still relevant even though the Financial Stability Board's Task Force on Climate-related Financial Disclosures recently published its latest TCFD status report in October 2023 (TCFD 2023) where they announced that "the ISSB standards represent a culmination of the Task Force's work and that the TCFD would be disbanded upon release of its 2023 status report" (TCFD 2023, ii).

[3] It should be noted that the first Carbon Disclosure framework was developed by the CDP (Carbon Disclosure Project) which was established in 2000. In its recent status report, the TCFD notes that "earlier this year, CDP announced that over 680 financial institutions with more than $130 trillion in assets have asked over 10,000 companies to disclose through CDP, which has aligned its climate change disclosures with the TCFD recommendations" (FSB-TCFD 2022, 3). I discuss CDP scores and their methodology in later sections.

[4] Emphasis added.

Opportunities, and Financial Impact (see Papazian 2022, for a detailed discussion).

Three years after the publication of the TCFD framework, the TNFD or the Taskforce on Nature-related Financial Disclosures was announced in July 2020. Its purpose and objective were described to be the delivery of "a risk management and disclosure framework for organisations to report and act on nature-related risks" (TNFD 2021a). In March 2023, the TNFD published its own framework, 'The TNFD Nature-related Risk and Opportunity Management and Disclosure Framework' (TNFD 2023a).[5]

The report states that the TNFD framework is strategically, and wherever relevant, technically aligned with the TCFD framework. Indeed, in the report's own words, it "retained the four pillars of the TCFD Recommendations – Governance, Strategy, Risk Management and Metrics and Targets – with Impact Management incorporated into Risk Management" (TNFD 2023, 14).[6]

Since its publication, the TCFD framework has been further developed and refined, and a series of publications by the TCFD Portfolio Alignment Team (TCFD-PAT) propose a number of methodologies to align portfolios with relevant pathways that would lead to the achievement of our climate targets. The next subsection discusses one of the key proposed metrics, the Implied Temperature Rise (ITR).

3.3.1 Portfolio Alignment and Implied Temperature Rise

The TCFD report by the Portfolio Alignment Team (TCFD-PAT 2021) identifies a number of approaches that could be used in the alignment of portfolios with our climate targets. The simplest is based on measuring alignment by the percentage of investments in a portfolio with declared

---

[5] Note that TNFD and TCFD, although similar in purpose and acronym, do not have the same institutional background. The TCFD was created by the Financial Stability Board (FSB) following a request by the G20, while the TNFD was created by a group of financial institutions and corporations, and it is not formally linked to the FSB. It does receive support and funding from a number of institutions like the UNDP, and public agencies from a variety of countries such as Australia, France, Germany, and the UK.

[6] "The Taskforce is made up of 40 senior executives from corporates, financial institutions and market intermediaries from around the world.... Collectively, the Taskforce members represent institutions with over US$20 trillion in assets under management and a footprint in over 180 countries across five continents" (TNFD 2023a, 5).

Net Zero/Paris alignment targets. A very basic assessment that does not explore the targets' reliability, effectiveness, or implementation.

Another approach is to use 'benchmark divergence models,' which measure alignment at each investment level by comparing asset emissions with a benchmark emission pathway based on forward-looking scenarios. Given that mitigation pathways and forward-looking scenarios are themselves a tentative collection of many policy and industry-level transformations, actions, and conditions, this approach is slightly more complex, but does not necessarily provide a comparable assessment given the diversity of pathways and scenarios assessing institutions can use.

Last but not least, the Implied Temperature Rise model (ITR), a more complex approach, aims to measure the potential global warming outcome if the economy was to have the same level of emissions as the asset or portfolio under consideration. It is defined as follows:

> Implied temperature rise (ITR) models: These tools extend benchmark divergence models one step further, translating an assessment of alignment/misalignment with a benchmark into a measure of the consequences of that alignment in the form of a temperature score that describes the most likely global warming outcome if the global economy was to exhibit same level of ambition as the counterparty in question. (TCFD-PAT 2021, 2)

The main suggestion of the TCFD-PAT report is that institutions "use whichever portfolio alignment tool best suits their institutional context and capabilities." Loosely termed and with many moving parts, portfolio alignment is left at the discretion and capabilities of the institutions doing their own assessments.

The Cambridge Institute for Sustainability Leadership—Investment Leaders Group published a detailed methodological paper on temperature scores (ITR) describing a four-step process (CISL 2021a, b). I discuss the CISL (2021b) approach here, given the revealing equations, and despite a fundamental shortcoming in the model. The below discussion was first published in Papazian (2022, 64–67).

The first step of the process as described by CISL is the measurement of the carbon intensity of the asset/portfolio, then the calculation of its equivalent emissions, then its cumulative emissions, and finally its temperature score.

1. We calculate the ITR measure using the CISL (2021b) method by first calculating the carbon intensity of the asset:

$$\text{CERI Asset} = \frac{\text{Scope 1 and 2 Emissions}}{\text{Revenue}}$$

2. We calculate the equivalent emissions that calculates the global emissions if the world had the same carbon intensity as the asset:

$$\text{Equivalent Global Emissions} = \text{CERI}_{\text{Asset}} * \text{Global GDP} * \theta$$

where Theta is a scaling factor equal to 2.61, calculated through:

$$\theta = \frac{\text{Global benchmark for emissions intensity}}{\text{Portfolio benchmark for emissions intensity}}$$
$$= \frac{493.18\,[\text{tCO2/US\$m}]}{188.70\,[\text{tCO2/US\$m}]} = 2.61$$

3. We calculate the equivalent global cumulative emissions of the asset:

$$\text{Cumulative CO}_2 \text{ Emissions} = \sum_{t=2020}^{2100} \text{Equivalent Global Emissions}_t$$

4. Using the transient climate response to cumulative CO_2 emissions (TCRE) formula to derive the temperature score,

$$\text{Global Warming Since 2020} = \alpha * \text{cumulative CO}_2 \text{ Emissions Since 2020} + \beta$$

where $\alpha = 5.29 \cdot 10^{-4}$ ($°C/GtCO2$) and $\beta = 1.24$ ($°C$), we find the implied temperature score of this investment.

When looking for the implied temperature rise score of a portfolio, the logic remains the same, but the calculation of CERI differs. It is equal to:

$$\text{CERI}_{\text{Portfolio}} = \frac{\sum_{i=1}^{n \text{ assets}} \left(\frac{\text{Value of Investment}_i}{\text{EVIC}_i} \times \text{Scope 1 and 2 GHG Emissions}_i \right)}{\sum_{i=1}^{n \text{ assets}} \left(\frac{\text{Value of Investment}_i}{\text{EVIC}_i} \times \text{Revenue}_i \right)}$$

where Value of Investment is the proportion of the portfolio invested in asset i, and EVIC is the Enterprise Value including Cash of the asset i.

$$\text{Value of Investment}_i = \text{Size of Fund} * \text{Weight of Asset}_i$$

Enterprise Value including Cash is commonly defined as:

$$EVIC_i = \text{Common Shares} + \text{Preferred Shares}$$
$$+ \text{Market Value of Debt} + \text{Minority Interest}$$

The above equations reveal that ITR does *not* change or affect the valuation of the asset; it does *not* change or affect its enterprise value; it simply measures its relative performance in temperature terms using a series of assumptions. As I have argued previously, this model, through the scaling factor Theta (θ), penalises the climate reporting assets in the portfolio given the non-reporting firms in the index used to calculate Theta (CISL 2021b, 66).

> The denominator in the Theta fraction is low, thus leading to a higher Theta, because around 40% of the companies in the index used for the calculation do not yet report emissions. Assuming that they are potentially the ones with much higher intensity, this model is actually ascribing much higher scores to reporting companies simply because of the non-reporting firms, a bias that can send very damaging market signals. Furthermore, it is not clear why this scaling factor has been used, as the model is built on the assumption that it calculates the resulting equivalent emissions if the world economy were to have the same carbon intensity of the asset or the portfolio. Multiplying that figure by 2.61 is not clearly justified.
>
> Furthermore, not including scope 3 emissions due to data availability is understandable, but from a modelling perspective, it reinforces the issue discussed above regarding the scaling factor used, Theta. If the scope three emissions of the companies in the index were to be included, alongside the emissions of those who do not report yet, Theta will most likely be equal to one. (Papazian 2022, 66)

The concept/metric of Implied Temperature Rise (ITR) may be interesting to consider, but in itself, it does not change nor transform the equations of value and return being applied to the assets and/or portfolios. Moreover, ITRs do not measure actual impact, and do not change or affect the valuation of the asset or portfolio under consideration. ITRs may hypothetically identify a *misaligned asset or portfolio*, but they do not describe a realignment, transformation, and transition logic.

To conclude this section, the main framework that has come to garner global support within the finance industry, TCFD (2017, 2022), and the parallel framework on nature (TNFD 2023a), and the key portfolio alignment tool, the Implied Temperature Rise, besides being heavily risk-oriented, they do not go far enough to transform our equations of value and return.

3.4   ESG Ratings

It is a self-evident fact that ESG factors (Environmental, Social, and Governance) have grown in importance. This has been due to the development of ESG-linked frameworks, but also to investor attention (Serafeim 2020). Although, many larger firms develop and use their own proprietary ESG scores, one of the most commonly used tools through which investors have taken ESG factors into account are ESG ratings.

ESG ratings/scores are used by market participants, such as asset managers, to assess the compliance and performance of their investments. There is a growing academic literature in finance that looks at ESG and seeks to find evidence of their relevance to risk, value, and asset prices (Capelli et al. 2023; Berg et al. 2022; Christensen et al. 2021; Avramov et al. 2021; Gibson et al. 2021; Cornell and Damodaran 2020; Cornell and Shapiro 2021; Sherwood and Pollard 2019; Inderst and Stewart 2018; Friede et al. 2015). Before discussing some of those findings, it is important to understand what ESG ratings actually measure and how.

3.4.1   Ratings Methodology

Table 3.1 presents the scores, grades, and approach used by Refinitiv (2021), one of the main ESG ratings providers in the market. The report that summarises the methodology (Refinitiv 2021) describes the pillar scores (E, S, and G), pillar weights, categories, materiality matrix, and the type of data points used to assess what is called the 'controversy overlay score.' 21 out of 23 data points used for the ESG controversy score start with "number of controversies published in the media..." (Refinitiv 2021, 23). The media is used as a data source in order to assess the reporting companies' actions against commitment. In other words, any non-compliance or abuse of the environment that does not make it into the media, is missed.

In a note on Boolean and numeric data points, the report states:

Table 3.1 Refinitiv ESG scores, grades, scoring method

Score range	Grade	Description
$0.0 \leq$ score ≤ 0.083333	D−	'D' score indicates poor relative ESG
$0.083333 <$ score ≤ 0.166666	D−	performance and insufficient degree of
$0.166666 <$ score ≤ 0.250000	D+	transparency in reporting material ESG data publicly
$0.250000 <$ score ≤ 0.333333	C−	'C' score indicates satisfactory relative
$0.333333 <$ score ≤ 0.416666	C−	ESG performance and moderate degree
$0.416666 <$ score ≤ 0.500000	C+	of transparency in reporting material ESG data publicly
$0.500000 <$ score ≤ 0.583333	B−	'B' score indicates good relative ESG
$0.583333 <$ score ≤ 0.666666	B−	performance and above average degree
$0.666666 <$ score ≤ 0.750000	B+	of transparency in reporting material ESG data publicly
$0.750000 <$ score ≤ 0.833333	A−	'A' score indicates excellent relative
$0.833333 <$ score ≤ 0.916666	A−	ESG performance and high degree of
$0.916666 <$ score ≤ 1	A+	transparency in reporting material ESG data publicly

The scores are based on relative performance of ESG factors with the company's sector (for environmental and social) and country of incorporation (for governance). Refinitiv does not presume to define what 'good' looks like; we let the data determine industry-based relative performance within the construct of our criteria and data model. Refinitiv's ESG scoring methodology has a number of key calculation principles set out below

1. Unique ESG magnitude (materiality) weightings have been included—as the importance of ESG factors differs across industries, we have mapped each metric's materiality for each industry on a scale of 1–10
2. Transparency stimulation—company disclosure is at the core of our methodology. With applied weighting, not reporting 'immaterial' data points doesn't greatly affect a company's score, whereas not reporting on 'highly material' data points will negatively affect a company's score
3. ESG controversies overlay—we verify companies' actions against commitments, to magnify the impact of significant controversies on the overall ESG scoring. The scoring methodology aims to address the market cap bias from which large companies suffer by introducing severity weights, which ensure controversy scores are adjusted based on a company's size
4. Industry and country benchmarks at the data point scoring level—to facilitate comparable analysis within peer groups
5. Percentile rank scoring methodology—to eliminate hidden layers of calculations. This methodology enables Refinitiv to produce a score between 0 and 100, as well as easy-to-understand letter grades (Refinitiv 2021, 3)

Source Refinitiv (2021, 7)

For instance, the answer to 'Does the company have a water efficiency policy?' can be 'Yes' (which is equal to 1) if this is indeed the case, or 'No' if the company in question does not have such a policy, or if it reports only partial information (which is equivalent to 'No'). (Refinitiv 2021, 9)

As it is clear from the above, ESG ratings do not measure impact. Notice in Table 3.1 how the grades refer to '*relative* ESG performance and ... degree of transparency in reporting material ESG data publicly.' They are more like scores on reporting than actual impact.

Similarly, S&P Global reveals (S&P Global 2022b) a questionnaire-based scoring system. The 'Corporate Sustainability Assessment' is an annual assessment with 61 industry-specific questionnaires. When companies do not provide answers, the S&P team complete them according to their own research.

> ESG Scores are measured on a scale of 0 – 100, where 100 represents the maximum score. Points are awarded at the question-level (on average 130 per company) based on our assessment of underlying data points (up to 1,000 per company) according to pre-defined scoring frameworks that assess their availability, quality, relevance, and performance on ESG topics. (S&P Global 2022b, 3)

In Appendix I of the same report, we find an example of the calculation process:

> In 2021, the environmentally oriented bank, Blue Sycamore Bank, participated in the S&P Global Corporate Sustainability Assessment (CSA). Due to their comprehensive environmental impact reporting practices, they received 100(/100) points on the 'Environmental Reporting – Coverage' *question*. Within the 'Environmental Reporting' *criteria* for the 2021 Banks CSA, the 'Environmental Reporting – Coverage' question has a weight of *50%*. Therefore, the 'Environmental Reporting – Coverage' 100 question points contributed 50 points (100 * 0.5) to the 'Environmental Reporting' criteria. Within the 'Environmental' *dimension* for the 2021 Banks CSA, the 'Environmental Reporting' criteria has a weight of *23%*. Therefore, the 'Environmental Reporting – Coverage' 100 question points contributed 11.5 points ((100 * 0.5) * (0.23)) to the 'Environmental' dimension. Within the S&P Global *ESG Score* for the 2021 Banks CSA,

the 'Environmental' dimension has a weight of *13%*. Therefore, the 'Environmental Reporting – Coverage' 100 question points contributed 1.5 points ((100 * 0.5) * (0.23) * (0.13)) to the S&P Global ESG Score. (S&P Global 2022b, 13)[7]

The score and weights of each aggregation level, i.e., Question, Criteria, Dimension, ESG Score, and the number of initial questions and data points, i.e., 130 and 1000 per company, reveal that the final score is a subjective summary of many different small and large, significant and not so significant data points, without investigating or measuring actual impacts.

On the 31st of July 2023, a *Financial Times* article titled "Companies with good ESG scores pollute as much as low-rated rivals" (Johnson 2023) refers to a research done by Scientific Beta (2021), and quotes its director, Felix Goltz:

> ESG ratings have little to no relation to carbon intensity, even when considering only the environmental pillar of these ratings... It doesn't seem that people have actually looked at [the correlations]. They are surprisingly low. (Johnson 2023)

In truth, a number of studies have looked at ESG ratings and their correlations to emissions. Boffo et al. 2020, Boffo and Patalano (2020) have addressed this issue in an OECD study in 2020.

Boffo et al. (2020) state:

> [W]hile the E score includes a number of distinct environmental metrics, the analysis found a positive correlation between some ESG raters' high E scores of corporate issuers and high levels of carbon emissions and waste. (Boffo et al. 2020, 7)

Previous research has also explored correlations of ESG ratings from a variety of perspectives. One key finding has revealed that there is a 'disagreement' between popular ESG ratings in the market and speak of 'rater effect' (Berg et al. 2022; Christensen et al. 2021). Berg et al. (2022) reveal ESG ratings divergence amongst the top six providers, i.e., KLD, Sustainalytics, Moody's ESG, S&P Global, Refinitiv, and MSCI,

---

[7] Emphases added to denote the layers of aggregation and weights being applied.

raising the necessity to look at "how the data underlying ESG ratings is generated."

Indeed, ESG ratings do not necessarily correlate as expected with the emissions of the companies concerned, and they often diverge from each other. As I have argued in Papazian (2022), they are more like opinion points rather than data points—not comparable, and their methodologies are often shrouded in commercial strategic vagueness.

In a recent study, Gibson et al. (2021) found that stock returns are positively related to ESG rating disagreement, and that this is mainly based on disagreement about the environmental aspect of the scores. Based on the evidenced divergence or uncertainty of ESG ratings, studies have also investigated their relationship to performance and returns (Avramov et al. 2021).

The evidence is mixed and inconclusive when it comes to ESG and returns. Cornell and Damodaran (2020) summarise the issue eloquently:

> In many circles ESG is being marketed as not only good for society, but good for companies and for investors. In our view, however, the hype regarding ESG has vastly outrun the reality of both what it is and what it can deliver. The potential to make money on ESG for consultants, bankers and investment managers has made them cheerleaders for the concept, with claims of the payoffs based on research that is ambiguous and inconclusive, if not outright inconsistent with some of the claims. (Cornell and Damodaran 2020, 22)

Whatever their correlation with emissions, corporate performance, and asset prices, ESG ratings or scores do not measure impact. Moreover, besides being infrequent, they also do not provide an interpretive framework through which markets can interpret marginal new information that may affect the scores (Papazian 2022).

To conclude this section, ESG ratings do not transform our equations of value, and they do not transform our value framework in finance. They do not measure impact, and they do not integrate the impact of cash flows into the valuation of those cash flows.

3.4.2 ESG Integration in Practice

In 2016, the Principles of Responsible Investment (PRI 2016) in collaboration with the United Nations Environment Program Finance Initiative

(UNEPFI) and United Nations Global Compact (UNGC) published a report that discusses ESG integration in practice with direct contributions from asset managers. The report focused on listed equities.

The report defines ESG integration as:

> The PRI defines ESG integration as "the systematic and explicit inclusion of material ESG factors into investment analysis and investment decisions". It is one of three ways to incorporate responsible investment into investment decisions, alongside thematic investing and screening. All three ESG incorporation practices can be applied concurrently. (PRI 2016, 12)

I have discussed the 2016 report in Papazian (2022), the below paragraph summarises the main conclusion.

> The ESG integration strategies and examples often use clip-art tables and drawings to demonstrate the many aspects of E and S and G and how they should be integrated, but in the end, the value equations and the value paradigm ... remain intact. Instead, ESG integration is tantamount to adjusting conventional valuation variables in conventional valuation models. (Papazian 2022, 73)

Indeed, the report states:

> Forecasted company financials drive valuation models such as the discounted cash flow (DCF) model, which in turn calculates the estimated value (or fair value) of a company and hence can affect investment decisions. Investors can adjust forecasted financials such as revenue, operating cost, asset book value and capital expenditure for the expected impact of ESG factors. (PRI 2016, 13)

The financial valuation models themselves remain intact. The Discounted Cash Flow (DCF) model remains central in calculating the fair value of a stock or cash flows. The relevance of ESG factors is determined through their influence on the forecasted variables used in the DCF models.

In a joint report published by PRI and the CFA Institute (2018), the process of ESG integration is described and developed in a 177-page document. PRI-CFA (2018) introduces an in-depth ESG integration framework built on three 'circles,' inner, middle, and outer. Each of them defines a set of conceptual, organisational, and structural requirements for the process, covering equity, fixed income, and portfolio-level analysis. The inner circle covers research, the middle circle covers security valuation, and the outer circle the portfolio-level analysis.

The report states that "[t]he ESG Integration Framework is not meant to illustrate the perfect ESG-integrated investment process. Rather, the ESG Integration Framework is meant to be a reference so that practitioners can analyze their peers' ESG integration techniques and identify those techniques that are suitable for their own firm" (PRI-CFA, 2018, 6). Looking at the details of what ESG integration implies for the valuation of securities, the middle circle, the report describes each element in detail. I have quoted the text in a table format verbatim from PRI-CFA (2018, 7–8) in Table 3.2. The adjustments described are unambiguous. They all concern variables and elements used in conventional financial forecasting, such as revenues, operating costs, asset book values, capital expenditures, discount rates, perpetuity growth rates, terminal values, valuation multiples, maturities, credit assessments, etc.

In 2023, PRI released its latest report on the subject (PRI 2023) where it defines ESG integration as follows:

> The PRI defines ESG integration as "including ESG factors in investment analysis and decisions *to better manage risks and improve returns.*" This means that investors assess ESG risks and opportunities when deciding whether or not to buy a stock… It is one of three ways to incorporate ESG considerations into investment processes, alongside thematic investing and screening. All three practices can be applied concurrently. (PRI 2023, 16)[8]

The focus on *managing risks and improving returns* is made even clearer in the 2023 report. ESG integration is defined as a five-part process including: (1) RI policy and beliefs, (2) governance, (3) investment process, (4) stewardship, and (5) monitoring and reporting. PRI (2023) is far more detailed and wider in depth and breadth, however, the key proposition of ESG integration remains the same. The process does not involve a reinvention of our value models used by asset managers, but rather, the consideration of ESG factors in the process of forecasting the future expected cash flows, terminal value, and current intrinsic value of the stock.

---

[8] Emphasis added.

Table 3.2 Security level—the middle circle

Security level—the middle circle	Description
Security valuation—equities	
Forecasted financials	Adjustments are made to forecasted financials (e.g., revenue, operating cost, asset book value, capital expenditure) for the expected impact of ESG factors
Valuation-model variables	Adjustments are made to valuation-model variables (e.g., discount rates, perpetuity growth, terminal value) for the expected impact of ESG factors
Valuation multiples	Adjustments are made to valuation multiples to calculate "ESG-integrated" valuation multiples. These multiples are then used to calculate the value of securities
Forecasted financial ratios	Forecasted financials and future cash flow estimates are adjusted for ESG analysis and the effect on financial ratios is assessed
Security sensitivity/scenario analysis	Adjustments are made to variables (sensitivity analysis) and different ESG scenarios (scenario analysis) are applied to valuation models to compare the difference between the base-case security valuation and the ESG-integrated security valuation
Security valuation—fixed income	
Credit analysis	– Internal credit assessments: ESG analysis is used to adjust the internal credit assessments of issuers – Forecasted financials and ratios: Forecasted financials and future cash flow estimates are adjusted for ESG analysis and the effect on financial ratios is assessed – Relative ranking: ESG analysis impacts the ranking of an issuer relative to a chosen peer group
Relative value analysis/spread analysis	An issuer's ESG bond spreads and its relative value versus those of its sector peers are analysed to find out if all risk factors are priced in

(continued)

Table 3.2 (continued)

Security level—the middle circle	Description
Duration analysis	The impact of ESG issues on bonds of an issuer with different durations/maturities is analysed
Security sensitivity/scenario analysis	Adjustments to variables (sensitivity analysis) and different ESG scenarios (scenario analysis) are applied to valuation models to compare the difference between the base-case security valuation and the ESG-integrated security valuation

Source PRI-CFA (2018, 7–8)

Discussing ESG integration in fundamental investment strategies, the report states:

> Traditionally, many investment managers would forecast the company financials (or at least establish views on likely future paths) and subsequently adjust those to reflect a set of ESG considerations.... But a more thorough integration of ESG factors would take these into account when forecasting sales, earnings and cash flows in the first place; rather than applying ESG adjustments afterward. (PRI 2023, 28)

As stated plainly and clearly, the forecasted variables are very much the same: sales, earnings, and cash flows, etc., and ESG factors are considered before or after or during the process of conventional forecasting.

We can see clearly that ESG integration is a process through which investors consider Environmental, Social, and Governance factors, and the end result of that process is *a series of adjustments to variables in our existing models.*

ESG integration, whether done through proprietary scores or market-purchased ratings, does not lead to a set of transformed value and return equations. There is no real change in the value framework and equations through which the value of a security is gauged.

3.4.3   CDP Scores and Methodology

The Carbon Disclosure Project (CDP) is one of the most popular carbon and environmental reporting frameworks. A non-profit organisation founded in 2000, CDP's latest brochure states that more than "18,700+ companies worth over half of global market capital, 740+ financial institutions worth US$130+ trillion in assets disclose through CDP" (CDP 2023b, 2). Also referred to in the TCFD (2022) annual status report, CDP has its own scoring methodology and an Online Response System (ORS) to collect the answers of reporting organisations.

CDP publishes its scores on an annual basis for a fee. In CDP (2023a), we can find an overview of the methodology used to calculate the scores. In the disclaimer of the document, we can read:

> The CDP score is based on activities and positions disclosed in the CDP response. It therefore does not consider actions not mentioned in the CDP response and data users are asked to be mindful that these may be positive or adverse or negative in terms of environmental management. *The score is not a comprehensive metric of a company's level of sustainability or 'green-ness', or a specific metric on the environmental footprint, but rather an indication of the level of action taken by the company to assess and manage its impacts on, and from, environmental related issues during the reporting year.* (CDP 2023a, 16)[9]

CDP claims that its scores provide 'a snapshot of a company's disclosure and environmental performance,' and that its methodology is aligned with the Taskforce for Climate-Related Financial Disclosures (TCFD) and with other major environmental standards. Expanding on the methodology of point allocation, the report states:

> Responding companies will be assessed across four consecutive levels which represent the steps a company moves through as it progresses towards environmental stewardship. The levels are: 1) Disclosure; 2) Awareness; 3) Management; 4) Leadership. At the end of scoring, the number of points a company has been awarded at Disclosure and Awareness level is divided by the maximum number that could have been awarded. The fraction is then converted to a percentage by multiplying by 100, working to two decimal places. For Management and Leadership levels, the number of

---

[9] Emphasis added.

points achieved per scoring category are used to calculate the final score, according to the scoring category weighting. (CDP 2023a, 6).

In other words, CDP scores, though popular, do not affect nor influence our core finance principles and equations of value and return. They, as mentioned above (CDP 2023a, b, 16), are not "a specific metric on the environmental footprint" of the reporting entities.

3.5   IMPACT INVESTING AND MEASUREMENT

In parallel to ESG ratings and ESG integration, one of the most prominent strands of sustainable finance is impact investing (Bugg-Levine and Emerson 2011; Jackson 2013; Bertl 2016; Reeder et al. 2015; Nusseibeh 2017; Agrawal and Hockerts 2018). A recent study by the Impact Investing Institute revealed that the UK impact investment market was worth nearly £58 billion in 2020, which is estimated to represent around 3.3–8% of the total global market (Impact II 2022). The definition used by the institute is adopted from the Global Impact Investing Network (GIIN), and it defines impact investing as follows: "impact investments are investments made with the intention to generate positive, measurable social and environmental impact alongside a financial return" (GIIN 2023).

At the heart of the impact investing field is the definition and measurement of impact (Impact Institute 2019). A variety of approaches to the measurement of impact have been in development. The conceptualisation of impact is often done through different concepts of 'capital,' like the five capitals framework proposed by the Forum for the Future (FFTF 2011), the six capitals framework proposed by the IIRC (2021, 2013), and the three capitals inclusive wealth concept of UNEP (2018).

Writing more than a decade ago, Bugg-Levine and Emerson (2011) identify the challenge of a clear and replicable methodology of measurement of impact. While they propose the concept of 'blended value' combining the economic, social, and environmental components of value (2011, 10), they also state: "[d]espite this new attention, the fundamental challenge remains unresolved: How do we develop a measurement system that offers an integrated understanding of blended value creation that matches the interest of the impact investor?" (Bugg-Levine and Emerson 2011, 167).

The Global Impact Investing Network (GIIN 2016) discusses the different ways impact investors derive value from impact measurement and management (IMP 2016). While there are no equations for measurement, the report gives procedural advice on how investors can target and achieve the impact they desire given the risks and stakeholders involved. The GIIN refers to the Social Impact Investment Taskforce report titled 'Measuring Impact' for the definition of impact measurement, which in its 2014 report (SIIT 2014), defines the four key stages of impact measurement and management as: (1) plan, (2) do, (3) assess, and (4) review.

The Global Impact Investing Network (GIIN 2020) recognises the methodological and measurement challenge presented by impact:

> Yet gauging investment-level impact remains a challenge: until now, there has been no tested, widely accepted methodology to assess and, most critically, compare impact results. It may be easy to conclude that impact can and should be distilled into a monetary figure, yet impact is inherently multi-dimensional and complex. To optimize impact performance, investors must be able to differentiate the impact of one investment from the impact of another. Across the investment process, investors perceive opportunities to compare and optimize investments' potential or realized impact. (GIIN 2020, 2)

More recently, the Harvard Business School Impact-Weighted Accounts project (HBS-IWA 2022) was launched, building on the research done by Serafeim and Trinh (2020). The project aims "to drive the creation of financial accounts that reflect a company's financial, social, and environmental performance." More recently, the Value Balancing Alliance has put forward a number of methodological and framework papers on 'Impact Statements' (VBA 2021a, b, c). In March 2022, the Harvard Business School's Impact-Weighted Accounts initiative and the Value Balancing Alliance made a joint statement (HBS-VBA 2022) stating:

> Today, we recognize and welcome the international harmonization and standardization of frameworks for sustainability reporting and disclosures. However, the landscape of methodologies to assess corporate sustainability performance is still fragmented. To ensure robust and comparable information, efforts to standardize the definition, measurement, and valuation of positive and negative impacts from business on society and enterprise value need to be intensified.

Several methodological approaches are currently being explored. *Most fail to handle the measurement and integration of economic, social, and environmental aspects at the same time in a consistent way.* Harvard Business School Impact Weighted Accounts (HBS IWA) and the Value Balancing Alliance (VBA) are convinced that monetary valuation of impacts on society and enterprise value is the most promising way to further develop respective sustainability accounting systems in a meaningful, fast, and business compatible manner. (HBS-VBA 2022)[10]

It is beyond any doubt that the measurement of impact will continue to be work in progress in the near future. While many of the reports and institutions mentioned above have their own definitions and targets, with many frameworks and stakeholder flowcharts, "equations for actual impact measurement in monetary terms are still missing, and none offer any procedural recommendations through which impact can be integrated into our existing value equations in core finance theory and practice" (Papazian 2022, 77).

3.6   Space in Sustainable Finance

Following the discussion in Chapter 2, where the absence of the analytical dimension of space was clearly noted and observed in our core value principles and equations, it is important to note here that space is also missing in the frameworks, standards, and tools of sustainability.

Looking through the five key documents relating to our efforts of operationalising sustainability, i.e., TCFD (2017), GRI (2022), IFRS (2023a, b), and EU (2023c), we observe, as depicted in Table 3.3, that the most popular term in all these documents is 'risk.' This, as per our discussion in Chapter 2, should not come as a surprise. Our entire financial value framework is built around risk and time and it is natural that 'risk and time' would be at the very top.

What is surprising, is that 'space' is absent across the board, except in GRI (2022), where it is used referring to civic space (4), floor space (2), confined space (4), small space (1), and sacred space (2).[11] While commendable that such aspects of an entity and its operations are

---

10 Emphasis added.

11 Searching for the word 'responsibility,' we find it 6 times in the TCFD (2017) report, none of which refer to responsibility of impact, and 211 times in the GRI (2022) document. The main reason for this is the fact that the consolidated GRI standards

Table 3.3 Word count for risk, time, and space in key sustainability documents

	TCFD (2017)	GRI (2022)	IFRS (2023a) (S1)	IFRS (2023b) (S2)	EU (2023c) (ESRS)	Total
Risk	438	321	167	144	5	1075
Time	55	360	29	24	5	473
Space	0	13	0	0	0	13

Source Author

considered within the standards, this does not actually translate into the consideration of space as an analytical dimension, and it certainly does not lead to transformed value and return equations where space and space impact are considered and integral to our equations.

One important mention is due here regarding a recent initiative called 'spatial finance.' Spatial finance is defined as the integration of geospatial data in the management of risks, opportunities, and impacts related to the climate and the environment:

> Spatial Finance, the integration of geospatial data and analysis in financial theory and practice, allows financial institutions to understand and manage risks, opportunities and impacts related to climate and the environment in a granular and actionable way. Advancements in geospatial technologies

document includes numerous repeated references and disclaimers from subsections. On a closer look, we find that: 51 of the 211 instances come from the repeated reference or citation of the "United Nations, The Corporate Responsibility to Respect Human Rights: An Interpretive Guide, 2012." 80 of the 211 come from the repeated legal disclaimer: "the preparation and publication of reports based fully or partially on the GRI Standards and related Interpretations are the full responsibility of those producing them. Neither the GRI Board of Directors, GSSB, nor Stichting. Global Reporting Initiative (GRI) can assume responsibility for any consequences or damages resulting directly or indirectly from the use of the GRI Standards." Looking closer, we observe that a further 41 of them are the repeated description where GRI is referring to the responsibility of the standard being discussed: "Responsibility: This Standard is issued by the Global Sustainability Standards Board (GSSB)." From the remaining 39 instances of the use 'responsibility', we find a few more references, and texts in the guidelines about what the organisation must report: "Report the level and function within the organization that has been assigned responsibility for managing risks and opportunities due to climate change, and Report whether responsibility to manage climate change-related impacts is linked to performance assessments or incentive mechanisms." In other words, only a handful of times responsibility refers to responsibility for emissions or impact.

and data science are making it possible to collect asset-level information in a consistent, timely and scalable manner, increasing transparency for investors, policymakers, and civil society alike by aggregating insights at a company, sector or portfolio level from the physical asset level upwards. (CGFI-SFI 2021)

While an exciting and important initiative, spatial finance is defined by and focused on the use of geospatial data (SFPUO 2018; Caldecott 2019), rather than the transformation of our value paradigm with the introduction of the analytical dimension of space, our physical context, and the reinvention of our equations of value and return.

3.7   Conclusion

Following the discussion in Chapter 2, where we discussed our risk and time-based value framework, we observed that our current principles of value discriminate against our very own evolutionary investments. We also identified that our equations of value and return are missing the analytical dimension of space, and thus our responsibility for space impact.

This chapter explored the growing field of sustainable finance and the most prominent frameworks, standards, scores, and strategies used to operationalise sustainability in finance. The discussion revealed that sustainable finance is yet to penetrate the analytical content of core finance theory and equations.

Space, as an analytical dimension and our physical context, is still missing, and our equations of value remain unchanged. Moreover, many of the tools and metrics used in the field do not actually measure impact, nor do they transform our existing equations—those taught and applied in finance theory and practice, across academia and industry, and around the world.

Moreover, the review of ESG integration, the most common approach used by asset and investment managers, revealed that the practice is focused on using ESG factors to adjust the same old variables in our existing models and equations.

We also observed that across these frameworks, standards, scores, and strategies, sustainable finance is entirely focused on businesses, investments, and instruments, and seems to treat money and its creation as exogenous to the sustainability challenge. Money mechanics and our monetary architecture are absolved by way of omission.

In other words, the 'spaceless' and 'contextless' value framework that we observed in Chapter 2, which has governed our financial investments, capital allocation strategies, and money creation methodology, is still intact.

The next chapter explores a new framework, the space value framework (Papazian 2022), which offers one plausible approach where the standards of sustainability can be made to link to a new mathematics of value and return, where planet and humanity are equal stakeholders next to the mortal risk-averse investor, and where sustainability is hardwired into our models.

References

Adams, C.A., A. Alhamood, X. He, J. Tian, L. Wang, and Y. Wang. 2021. The Double-Materiality Concept Application and Issues. Global Reporting Initiative. https://www.globalreporting.org/media/jrbntbyv/griwhitepaper-public ations.pdf. Accessed 02 February 2022.

Agrawal, A., and K. Hockerts. 2018. Impact Investing: Review and Research Agenda. *Journal of Small Business & Entrepreneurship* 33 (2): 153–181. https://www.tandfonline.com/doi/full/10.1080/08276331.2018.155 1457. Accessed 02 February 2021.

Avramov, D., S. Cheng, A. Lioui, and A. Tarelli. 2021. Sustainable Investing with ESG Rating Uncertainty. *Journal of Financial Economics*. https://ssrn.com/ abstract=3711218 or https://doi.org/10.2139/ssrn.3711218. Accessed 02 February 2022.

Barker, R., and R.G. Eccles. 2018. Should FASB and IASB Be Responsible for Setting Standards for Nonfinancial Information? Said Business School, University of Oxford. https://www.sbs.ox.ac.uk/sites/default/files/ 2018-10/Green%20Paper_0.pdf. Accessed 02 February 2021.

Berg, F., J. Kölbel, and R. Rigobon. 2022. Aggregate Confusion: The Divergence of ESG Ratings. https://ssrn.com/abstract=3438533. Accessed 02 February 2022.

Bertl, C. 2016. Environmental Finance and Impact Investing: Status Quo and Future Research. *ACRN Oxford Journal of Finance and Risk Perspectives* 5 (2): 75–105. http://www.acrn-journals.eu/resources/jofrp0502f.pdf. Accessed 02 February 2021.

Boffo, R., and R. Patalano. 2020. ESG Investing: Practices, Progress and Challenges. Paris: OECD. https://www.oecd.org/finance/ESG-Investing-Practi ces-Progress-Challenges.pdf. Accessed 02 February 2022.

Boffo, R., C. Marshall, and R. Patalano. 2020. ESG Investing: Environmental Pillar Scoring and Reporting. Paris: OECD Paris. https://www.oecd.org/fin ance/esg-investing-environmental-pillar-scoring-and-reporting.pdf. Accessed 02 February 2022.

Bokern D. 2022. Reported Emission Footprints: The Challenge Is Real. MSCI Research. https://www.msci.com/www/blog-posts/reported-emission-footprints/03060866159. Accessed 11 March 2022.

Bugg-Levine, A., and J. Emerson. 2011. *Impact Investing*. San Francisco: Jossey-Bass.

Caldecott, B. 2019. Viewpoint: Spatial Finance has a key. IPE Magazine, https://www.ipe.com/viewpointspatial-finance-has-a-key-role-/10034269. article Accessed 02 February 2022.

Capelli, P., F. Ielasi, and A. Russo. 2023. Integrating ESG Risks into Value-at-Risk. *Finance Research Letters* 55 (Part A). https://doi.org/10.1016/j.frl. 2023.103875. Accessed 28 July 2023.

CBD. 2021. A New Global Framework for Managing Nature Through 2030. UN Convention on Biological Diversity. https://www.cbd.int/article/draft-1-global-biodiversity-framework. Accessed 02 February 2022.

CCC. 2019. Net Zero: The UK's Contribution to Stopping Global Warming. Climate Change Committee. http://www.theccc.org.uk/wp-content/upl oads/2019/05/Net-Zero-The-UKs-contribution-to-stopping-global-war ming.pdf. Accessed 12 June 2021.

CDP. 2020. Doubling Down: Europe's Low-Carbon Investment Opportunity. CDP Europe. https://cdn.cdp.net/cdp-production/cms/reports/docume nts/000/004/958/original/Doubling_down_Europe's_low_carbon_invest ment_opportunity.pdf. Accessed 12 December 2021.

CDP. 2023a. Scoring Introduction 2023. CDP. https://cdn.cdp.net/cdp-pro duction/cms/guidance_docs/pdfs/000/000/233/original/Scoring-Introd uction.pdf. Accessed 30 July 2023.

CDP. 2023b. Disclosing Through CDP: The Business Benefits. CDP. https:// cdn.cdp.net/cdp-production/comfy/cms/files/files/000/007/896/ori ginal/Benefits_of_Disclosure_brochure_2023.pdf. Accessed 30 July 2023.

CDP-WWF. 2020. Temperature Rating Methodology. CDP Worldwide and WWF International. https://cdn.cdp.net/cdp-production/comfy/cms/files/ files/000/003/741/original/Temperature_scoring_-_beta_methodology.pdf. Accessed 02 November 2021.

CGFI-SFI. 2021. State and Trends in Spatial Finance. Centre for Green Finance and Investment—Spatial Finance Initiative. https://www.cgfi.ac.uk/wp-con tent/uploads/2021/07/SpatialFinance_Report.pdf. Accessed 02 February 2022.

Christensen, D., G. Serafeim, and A. Sikochi. 2021. Why Is Corporate Virtue in the Eye of the Beholder? The Case of ESG Ratings. *The Accounting Review* 97: 147–175.

CISL. 2019. In Search of Impact Measuring the Full Value of capital. Cambridge Institute for Sustainability Leadership. https://www.cisl.cam.ac.uk/system/files/documents/in-search-of-impact-report-2019.pdf. Accessed 06 June 2021.

CISL. 2020. Measuring Business Impacts on Nature: A Framework to Support Better Stewardship of Biodiversity in Global Supply Chains. Cambridge: Cambridge Institute for Sustainability Leadership. https://www.cisl.cam.ac.uk/system/files/documents/measuring-business-impacts-on-nature.pdf. Accessed 02 February 2022.

CISL. 2021a. Understanding the Climate Performance of Investment Funds. Part 1: The Case for Universal Disclosure of Paris Alignment. Cambridge Institute for Sustainability Leadership. https://www.cisl.cam.ac.uk/system/files/documents/understanding-the-climate-performance-of.pdf. Accessed 02 February 2022.

CISL. 2021b. Understanding the Climate Performance of Investment Funds. Part 2: A Universal Temperature Score Method. Cambridge Institute for Sustainability Leadership. https://www.cisl.cam.ac.uk/download-understanding-climate-performance-investment-funds-part-2. Accessed 02 February 2022.

CPI. 2021. Global Landscape of Climate Finance 2021. Climate Policy Initiative. https://www.climatepolicyinitiative.org/wp-content/uploads/2021/10/Full-report-Global-Landscape-of-Climate-Finance-2021.pdf. Accessed 02 February 2022.

Cornell, B., and A.C, Shapiro. 2021. Corporate stakeholders, corporate valuation and ESG. *European Financial Management.* 27 (2): 196–207. https://doi.org/10.1111/eufm.12299. Accessed 29 July 2023.

Cornell, B., and A. Damodaran. 2020. Valuing ESG: Doing Good or Sounding Good?. NYU Stern School of Business, Available at SSRN: https://doi.org/10.2139/ssrn.3557432. Accessed 27 July 2023.

Dasgupta P. 2021. The Economics of Biodiversity: The Dasgupta Review. London: HM Treasury. https://assets.publishing.service.gov.uk/government/uploads/system/uploads/attachment_data/file/962785/The_Economics_of_Biodiversity_The_Dasgupta_Review_Full_Report.pdf. Accessed 02 March 2021.

EFRAG. 2021. Proposals for a Relevant and Dynamic EU Sustainability Reporting Standard-Setting. European Financial Reporting Advisory Group. https://www.efrag.org/Activities/2105191406363055/Sustainability-reporting-standards-interim-draft. Accessed 02 February 2022.

EU. 2018. Action Plan: Financing Sustainable Growth. European Commission. https://eur-lex.europa.eu/legal-content/EN/TXT/?uri=CELEX:52018D C0097. Accessed 02 February 2021.

EU. 2019. Guidelines on Non-financial Reporting: Supplement on Reporting Climate Related Information. https://eur-lex.europa.eu/legal-content/ EN/TXT/PDF/?uri=CELEX:52019XC0620(01)&from=EN. Accessed 02 February 2022.

EU. 2020a. Sustainable Finance Taxonomy—Regulation (EU) 2020/852. European Commission. https://ec.europa.eu/info/law/sustainable-finance-taxonomy-regulation-eu-2020-852_en. Accessed 02 February 2022.

EU. 2020b. Renewed Sustainable Finance Strategy and Implementation of the Action Plan on Financing Sustainable Growth. European commission. https://ec.europa.eu/info/publications/sustainable-finance-renewed-strategy_en. Accessed 02 February 2022.

EU. 2022. EU Taxonomy for Sustainable Activities. European Commission. https://ec.europa.eu/info/business-economy-euro/banking-and-finance/sustainable-finance/eu-taxonomy-sustainable-activities_en. Accessed 12 February 2022.

EU. 2023a. The Commission Adopts the European Sustainability Reporting Standards. European Commission. https://finance.ec.europa.eu/news/commission-adopts-european-sustainability-reporting-standards-2023-07-31_en. Accessed 31 July 2023.

EU. 2023b. Implementing and Delegated Acts—CSRD. European Commission. https://finance.ec.europa.eu/regulation-and-supervision/financial-services-legislation/implementing-and-delegated-acts/corporate-sustainability-reporting-directive_en. Accessed 31 July 2023.

EU. 2023c. Annex to Supplementing Directive 2013/34/EU of the European Parliament and of the Council as Regards Sustainability Reporting Standards. European Commission. https://ec.europa.eu/finance/docs/level-2-measures/csrd-delegated-act-2023-5303-annex-1_en.pdf. Accessed 31 July 2023.

EU TEG. 2019. On Climate Benchmarks and Benchmarks' ESG Disclosures. European Union Technical Expert Group. https://ec.europa.eu/info/sites/default/files/business_economy_euro/banking_and_finance/documents/190 930-sustainable-finance-teg-final-report-climate-benchmarks-and-disclosures_en.pdf. Accessed 06 June 2021.

FFTF. 2011. The Five Capitals—A Framework for Sustainability. Forum for the Future. https://www.forumforthefuture.org/the-five-capitals. Accessed 02 February 2021.

Friede, G., T. Busch, and A. Bassen. 2015. ESG and Financial Performance: Aggregated Evidence from More Than 2000 Empirical Studies. *Journal of Sustainable Finance & Investment* 5 (4): 210–233. https://www.tandfonline.com/doi/full/10.1080/20430795.2015.1118917. Accessed 02 February 2022.

Gibson, R., P. Krueger, and P. Schmidt. 2021. ESG Rating Disagreement and Stock Returns. *Financial Analyst Journal*. https://doi.org/10.1080/001 5198X.2021.1963186. Accessed 02 February 2022.

GFANZ. 2022. Financial Institution Net-Zero Transition Plans: Fundamentals, Recommendations, and Guidance. Glasgow Financial Alliance for Net Zero. https://assets.bbhub.io/company/sites/63/2022/09/Recommendations-and-Guidance-on-Financial-Institution-Net-zero-Transition-Plans-November-2022.pdf. Accessed 06 February 2023.

GFANZ. 2023. The Glasgow Financial Alliance for Net Zero. GFANZ. https://www.gfanzero.com/about/. Accessed July 2023.

GIIN. 2016. The Business Value of Impact Measurement. Global Impact Investing Network. https://thegiin.org/assets/GIIN_ImpactMeasurementReport_webfile.pdf. Accessed 24 December 2020.

GIIN. 2020. Methodology: For Standardizing and Comparing Impact Performance. Global Impact Investing Network. https://thegiin.org/assets/Methodology%20for%20Standardizing%20and%20Comparing%20Impact%20Performance_webfile.pdf. Accessed 02 February 2022.

GIIN. 2023. Impact Investing. The Global Impact Investing Network. https://thegiin.org/impact-investing/. Accessed 26 July 2023.

GRI. 2021. GRI Standards. Global Reporting Initiative. https://www.globalreporting.org/how-to-use-the-gri-standards/gri-standards-english-language/. Accessed 14 February 2022.

GRI. 2022. Consolidated Set of the GRI Standards. Global Reporting Initiative. https://www.globalreporting.org/how-to-use-the-gri-standards/gri-standards-english-language/. Accessed 30 July 2023.

HBS-IWA. 2022. Harvard Business School Impact Weighted Accounts. Harvard Business School. https://www.hbs.edu/impact-weighted-accounts/Pages/default.aspx. Accessed 02 March 2022.

HBS-VBA. 2022. Harvard Business School Impact-Weighted Accounts and Value Balancing Alliance Joint Statement. https://www.hbs.edu/impact-weighted-accounts/Documents/PI_HBS%20IWA-VBA_Joint%20Statement%20.pdf. Accessed 28 March 2022.

HM Government. 2019. Environmental Reporting Guidelines: Including Streamlined Energy and Carbon Reporting Guidance. The National Archives. https://assets.publishing.service.gov.uk/government/uploads/system/uploads/attachment_data/file/850130/Env-reporting-guidance_inc_SECR_31March.pdf. Accessed 06 June 2021.

HSBC-BCG. 2021. Delivering Net Zero Supply Chains: The Multi-Trillion Dollar Key to Beat Climate Change. Boston Consulting Group and HSBC. https://www.hsbc.com/-/files/hsbc/news-and-insight/2021/pdf/211026-delivering-net-zero-supply-chains.pdf?download=1. Accessed 02 February 2022.

IEA. 2017. Energy Technology Perspectives 2017: Catalysing Energy Technology Transformations. International Energy Agency. https://iea.blob.core.windows.net/assets/a6587f9f-e56c-4b1d-96e4-5a4da78f12fa/Energy_Technology_Perspectives_2017-PDF.pdf. Accessed 02 February 2021.

IEA. 2021. Net Zero by 2050: A Roadmap for the Global Energy Sector. International Energy Agency. https://iea.blob.core.windows.net/assets/deebef5d-0c34-4539-9d0c-10b13d840027/NetZeroby2050-ARoadmapfortheGlobalEnergySector_CORR.pdf. Accessed 02 February 2022.

IFRS. 2021. IFRS Foundation Announces International Sustainability Standards Board. https://www.ifrs.org/news-and-events/news/2021/11/ifrs-foundation-announces-issb-consolidation-with-cdsb-vrf-publication-of-prototypes/. Accessed 02 February 2022.

IFRS. 2023a. IFRS S1 General Requirements for Disclosure of Sustainability-Related Financial Information. IFRS. https://www.ifrs.org/issued-standards/ifrs-sustainability-standards-navigator/ifrs-s1-general-requirements/#standard. Accessed 30 July 2023.

IFRS. 2023b. IFRS S2 Climate-Related Disclosures. IFRS. https://www.ifrs.org/issued-standards/ifrs-sustainability-standards-navigator/ifrs-s2-climate-related-disclosures/#standard. Accessed 30 July 2023.

IIRC. 2013. Capitals—Background Paper for <IR>. The Technical Task Force of the International Integrated Reporting Council. The Technical Collaboration Group. https://www.integratedreporting.org/wp-content/uploads/2013/03/IR-Background-Paper-Capitals.pdf. Accessed 02 February 2021.

IIRC. 2021. International <IR> Framework. The International Integrated Reporting Council. https://www.integratedreporting.org/wp-content/uploads/2021/01/InternationalIntegratedReportingFramework.pdf. Accessed 02 February 2022.

Impact Institute. 2019. Framework for Impact Statements: Beta Version. The Impact Institute. https://www.impactinstitute.com/wp-content/uploads/2019/04/Framework-for-Impact-Statements-Beta-1.pdf. Accessed 02 July 2022.

Impact II. 2022. Estimating and Describing the UK Impact Investing Market. Impact Investing Institute. https://www.impactinvest.org.uk/wp-content/uploads/2023/04/Estimating-and-describing-the-UK-impact-investing-market.pdf. Accessed 20 June 2023.

Inderst, G., and F. Stewart. 2018. *Incorporating Environmental, Social, and Governance (ESG) Factors into Fixed Income Investment*. Washington: World Bank Group.

ILB, et al. 2020. The Alignment Cookbook—A Technical Review of Methodologies Assessing a Portfolio's Alignment with Low-Carbon Trajectories or Temperature Goal. Institut Louis Bachelier, Institute for Climate Economics, World Wildlife Fund, Ministere De La Transition Ecologique et Solidaire. https://www.institutlouisbachelier.org/wp-content/uploads/2021/03/the-alignment-cookbook-a-technical-review-of-methodologies-assessing-a-portfolios-alignment-with-low-carbon-trajectories-or-temperature-goal.pdf. Accessed 02 February 2022.

IMP. 2016. Impact Management Norms. Impact Management Project. https://impactmanagementproject.com/impact-management/impact-management-norms/. Accessed 02 February 2021.

IPBES. 2019. The Global Assessment Report on Biodiversity and Ecosystem Services. Intergovernmental Science-Policy Platform on Biodiversity and Ecosystem Services. https://ipbes.net/system/files/2021-06/2020%20IPBES%20GLOBAL%20REPORT%28FIRST%20PART%29_V3_SINGLE.pdf. Accessed 02 February 2021.

IPCC. 1988. IPCC History. Intergovernmental Panel on Climate Change. https://www.ipcc.ch/about/history/. Accessed 02 February 2022.

IPCC. 2013. Climate Change 2013: The Physical Science Basis. Summary for Policymakers. Intergovernmental Panel on Climate Change. https://www.ipcc.ch/site/assets/uploads/2018/03/WG1AR5_SummaryVolume_FINAL.pdf. Accessed 02 February 2021.

IPCC. 2018a. Summary for Policymakers. In Global Warming of 1.5°C. An IPCC Special Report on the Impacts of Global Warming of 1.5°C Above Pre-industrial Levels and Related Global Greenhouse Gas Emission Pathways, in the Context of Strengthening the Global Response to the Threat of Climate Change, Sustainable Development, and Efforts to Eradicate Poverty. IPCC. https://www.ipcc.ch/site/assets/uploads/sites/2/2019/05/SR15_SPM_version_report_LR.pdf. Accessed 02 February 2022.

IPCC. 2018b. Mitigation Pathways Compatible with 1.5°C in the Context of Sustainable Development. Intergovernmental Panel on Climate Change. https://www.ipcc.ch/site/assets/uploads/2018/11/sr15_chapter2.pdf. Accessed 02 February 2021.

IPCC. 2021. Climate Change 2021: The Physical Science Basis. IPCC. https://www.ipcc.ch/report/ar6/wg1/downloads/report/IPCC_AR6_WGI_SPM_final.pdf. Accessed 02 February 2022.

IPCC. 2022. Climate Change 2022: Impacts, Adaptation and Vulnerability. Summary for Policymakers. Intergovernmental Panel on Climate Change. https://report.ipcc.ch/ar6wg2/pdf/IPCC_AR6_WGII_SummaryForPolicy makers.pdf. Accessed 28 February 2022.

IPCC. 2023. Synthesis Report of the IPCC 6th Assessment Report (AR6): Summary for Policymakers. Intergovernmental Panel on Climate Change. https://report.ipcc.ch/ar6syr/pdf/IPCC_AR6_SYR_SPM.pdf. Accessed 22 March 2023.

Jackson, E.T. 2013. Interrogating the Theory of Change: Evaluating Impact Investing Where It Matters Most. *Journal of Sustainable Finance & Investment* 3 (2): 95–110. https://doi.org/10.1080/20430795.2013.776257. Accessed 02 February 2021.

Johnson, S. 2023. Companies with Good ESG Scores Pollute as Much as Low-Rated Rivals. *Financial Times.* https://www.ft.com/content/b95 82d62-cc6f-4b76-b0f9-5b37cf15dce4. Accessed 31 July 2023.

McKinsey GI. 2022. The Net-Zero Transition What It Would Cost, What It Could Bring. Mckensey Global Institute. https://www.mckinsey.com/~/media/mckinsey/business%20functions/sustainability/our%20insights/the%20net%20zero%20transition%20what%20it%20would%20cost%20what%20it%20could%20bring/the%20net-zero%20transition-report-january-2022-final.pdf. Accessed 14 February 2022.

MEA. 2003. Ecosystems and Human Well-Being: A Framework for Assessment. Millennium Ecosystem Assessment. Washington, DC: Island Press. http://pdf.wri.org/ecosystems_human_wellbeing.pdf. Accessed 02 February 2021.

MSCI. 2018. MSCI Carbon Footprint Index Ratios Methodology. MSCI. https://www.msci.com/documents/1296102/6174917/MSCI+Carbon+Footprint+Index+Ratio+Methodology.pdf/. Accessed 02 February 2022.

Nusseibeh, S. 2017. The "Why" Question in Investment Theory. Harvard Law School Forum on Corporate Governance. https://corpgov.law.harvard.edu/2017/03/28/the-why-question-in-investment-theory/. Accessed 10 March 2023.

Papazian, Armen. 2022. *The Space Value of Money: Rethinking Finance Beyond Risk and Time.* New York: Palgrave Macmillan. https://doi.org/10.1057/978-1-137-59489-1.

PRI. 2016. A Practical Guide to ESG Integration for Equity Investing. Principles of Responsible Investing. https://www.icgn.org/sites/default/files/2021-08/PRI_apracticalguidetoesgintegrationforequityinvesting.pdf. Accessed 30 July 2023.

PRI. 2020. PRI Annual Report. Principles of Responsible Investment. https://www.unpri.org/download?ac=10948. Accessed 06 June 2021.

PRI. 2023. ESG Integration in Listed Equity: A Technical Guide. Principles of Responsible Investing. https://www.unpri.org/download?ac=18407. Accessed 30 July 2023.

PRI-CFA. 2018. Guidance and Case Studies for ESG Integration: Equities and Fixed Income. Principles of Responsible Investment and CFA Institute. https://www.unpri.org/download?ac=5962. Accessed 30 July 2023.

Reeder, N., A. Colantonio, J. Loder, and G.R. Jones. 2015. Measuring Impact in Impact Investing: An Analysis of the Predominant Strength That Is Also Its Greatest Weakness. *Journal of Sustainable Finance & Investment* 5 (3): 136–154. https://www.tandfonline.com/doi/full/10.1080/20430795.2015.1063977. Accessed 02 February 2021.

Refinitiv. 2021. Environmental, Social, and Governance Scores from Refinitiv. Refintiv. https://www.refinitiv.com/content/dam/marketing/en_us/documents/methodology/refinitiv-esg-scores-methodology.pdf. Accessed 02 February 2022.

SASB. 2020a. SASB Implementation Supplement: Greenhouse Gas Emissions and SASB Standards. Sustainability Accounting Standards Board. https://www.sasb.org/wp-content/uploads/2020/10/GHG-Emmissions-100520.pdf. Accessed 02 February 2022.

SASB. 2020b. SASB Human Capital Bulletin. Sustainability Accounting Standards Board. https://www.sasb.org/wp-content/uploads/2020/12/HumanCapitalBulletin-112320.pdf. Accessed 02 February 2022.

SASB. 2020c. Proposed Changes to the SASB Conceptual Framework & Rules of Procedure—Bases for Conclusions and Invitation to Comment. Sustainability Accounting Standards Board. https://www.sasb.org/wp-content/uploads/2021/07/PCP-package_vF.pdf. Accessed 02 February 2022.

SASB. 2021. SASB Standards. Sustainability Accounting Standards Board. https://www.sasb.org/standards/download/. Accessed 02 February 2022.

SEBTi. 2021. Business Ambition for 1.5°C: Responding to the Climate Crisis. Science Based Targets Initiative. https://sciencebasedtargets.org/resources/files/status-report-Business-Ambition-for-1-5C-campaign.pdf. Accessed 02 February 2022.

SEBTi. 2022. Financial Sector Science-Based Targets Guidance. Science Based Targets Initiative. https://sciencebasedtargets.org/resources/files/Financial-Sector-Science-Based-Targets-Guidance.pdf. Accessed 14 February 2022.

SEC. 2022. SEC Proposes Rules to Enhance and Standardize Climate-Related Disclosures for Investors. US Securities and Exchange Commission. https://www.sec.gov/news/press-release/2022-46. Accessed 21 March 2022.

SEEA. 2014. System of Environmental Economic Accounting 2012—Central Framework. United Nations, European Commission, International Monetary Fund, The World Bank, OECD, FAO. https://unstats.un.org/unsd/envaccounting/seeaRev/SEEA_CF_Final_en.pdf. Accessed 02 February 2021.

Serafeim, G. 2020. Social-Impact Efforts That Create Real Value. *Harvard Business Review*. https://hbr.org/2020/09/social-impact-efforts-that-create-real-value. Accessed March 2023.

Serafeim, G., and K. Trinh. 2020. A Framework for Product Impact-Weighted Accounts. Harvard Business School Working Paper 20-076. https://www.hbs.edu/impact-weighted-accounts/Documents/Preliminary-Framework-for-Product-Impact-Weighted-Accounts.pdf. Accessed 02 February 2022.

SFPUO. 2018. Climate Risk Analysis from space: remote sensing, machine learning, and the future of measuring climate-related risk. Sustainable Finance Programme. https://www.smithschool.ox.ac.uk/research/sustainable-finance/publications/Remote-sensing-data-and-machine-learning-in-climate-risk-analysis.pdf. Accessed 02 February 2021.

Sherwood, W.M., and J. Pollard. 2019. *Responsible Investing: An Introduction to Environmental, Social and Governance Investments*. London and New York: Routledge.

SIIT. 2014. Measuring Impact. The Social Impact Investment Taskforce. https://www.gov.uk/government/groups/social-impact-investment-tasforce or https://thegiin.org/research/publication/measuring-impact. Accessed 02 February 2022.

S&P Global. 2022a. The Sustainability Yearbook 2022. S&P Global. https://www.spglobal.com/esg/csa/yearbook/2022/downloads/spglobal_sustainability_yearbook_2022.pdf. Accessed 12 February 2022.

S&P Global. 2022b. S&P Global ESG Scores Methodology. Sustainable 1. S&P Global. https://www.spglobal.com/esg/documents/sp-global-esg-scores-methodology.pdf. Accessed 28 March 2022.

TCFD. 2017. Final Report: Recommendations of the Task Force on Climate-related Financial Disclosures. https://assets.bbhub.io/company/sites/60/2020/10/FINAL-2017-TCFD-Report-11052018.pdf. Accessed 02 February 2021.

TCFD. 2021a. Forward Looking Financial Metrics Consultation. Task Force on Climate-related Financial Disclosures. https://assets.bbhub.io/company/sites/60/2021/03/Summary-of-Forward-Looking-Financial-Metrics-Consultation.pdf. Accessed 02 February 2022.

TCFD. 2021b. Proposed Guidance on Climate-related Metrics, Targets, and Transition Plans. Task Force on Climate-related Financial Disclosures. https://assets.bbhub.io/company/sites/60/2021/05/2021-TCFD-Metrics_Targets_Guidance.pdf. Accessed 01 January 2022.

TCFD. 2022. Status Report: Task Force on Climate-related Financial Disclosures. https://assets.bbhub.io/company/sites/60/2022/10/2022-TCFD-Status-Report.pdf. Accessed 28 July 2023.

TCFD. 2023. Task Force on Climate-related Financial Disclosures: 2023 Status Report. Financial Stability Board. The Task Force on Climate-related Financial Disclosures. https://www.fsb.org/wp-content/uploads/P121023-2.pdf. Accessed 18 October 2023.

TCFD-PAT. 2020. Measuring Portfolio Alignment. https://www.tcfdhub.org/wp-content/uploads/2020/10/PAT-Report-20201109-Final.pdf. Accessed 02 February 2022.

TCFD-PAT. 2021. Measuring Portfolio Alignment: Technical Considerations. https://www.tcfdhub.org/wp-content/uploads/2021/10/PAT_Measuring_Portfolio_Alignment_Technical_Considerations.pdf. Accessed 12 March 2023.

TNFD. 2021a. Taskforce on Nature-related Financial Disclosures. https://tnfd.global/. Accessed 02 February 2022.

TNFD. 2021b. Nature in Scope: A Summary of the Proposed Scope, Governance, Work Plan, Communication and Resourcing Plan of the TNFD. Taskforce on Nature-related Financial Disclosures. https://tnfd.global/wp-content/uploads/2021/07/TNFD-Nature-in-Scope-2.pdf. Accessed 02 February 2022.

TNFD. 2021c. Proposed Technical Scope: Recommendations for the TNFD. Taskforce on Nature-related Financial Disclosures. https://tnfd.global/wp-content/uploads/2021/07/TNFD-%E2%80%93-Technical-Scope-3.pdf. Accessed 02 February 2022.

TNFD. 2023a. The TNFD Nature-related Risk and Opportunity Management and Disclosure Framework. Taskforce on Nature-related Financial Disclosures. https://framework.tnfd.global/wp-content/uploads/2023/03/23-23882-TNFD_v0.4_Integrated_Framework_v7.pdf. Accessed 28 June 2023.

TNFD. 2023b. TNFD Releases First Beta Version of Nature-related Risk Management Framework for Market Consultation. Taskforce on Nature-related Financial Disclosures. https://tnfd.global/news/tnfd-releases-first-beta-framework/. Accessed 16 March 2022.

TPT. 2023. Disclosure Framework. Transition Plan Taskforce. https://transitiontaskforce.net/wp-content/uploads/2023/10/TPT_Disclosure-framework-2023.pdf. Accessed 18 October 2023.

UBS-RI. 2022. The Responsible Investor ESG Yearbook 2022. Responsible Investor, UBS AG. https://www.esg-data.com/product-page/esg-yearbook-2022. Accessed 12 February 2022.

UN. 2016. The Sustainable Development Agenda. United Nations Sustainable Development Goals. https://www.un.org/sustainabledevelopment/development-agenda-retired/. Accessed 12 June 2020.

UNEP. 2018. Inclusive Wealth Report. United Nations Environment Program. https://wedocs.unep.org/bitstream/handle/20.500.11822/26776/Inclus ive_Wealth_ES.pdf. Accessed 02 February 2021.

UNFCCC. 2015. Paris Agreement. United Nations Framework Convention on Climate Change. https://unfccc.int/sites/default/files/english_paris_agre ement.pdf. Accessed 02 December 2020.

UNFCCC. 2020. Cities, Regions and Businesses Race to Zero Emissions. UNFCCC. https://unfccc.int/news/cities-regions-and-businesses-race-to-zero-emissions. Accessed 02 February 2021.

UNFCCC. 2021. COP 26 and the Glasgow Financial Alliance for Net Zero (GFANZ). https://racetozero.unfccc.int/wp-content/uploads/2021/04/GFANZ.pdf. Accessed 02 February 2022.

VBA. 2021a. VBA Disclosure Concept for Material Sustainability Matters. Value Balancing Alliance. https://www.value-balancing.com/_Resources/Persis tent/7/2/a/2/72a28deeed4e259bc414148b2660e631e0dfe3d3/VBA_Dis closure_Concept.pdf. Accessed 02 February 2022.

VBA. 2021b. Methodology—Impact Statement. Value Balancing Alliance. https://www.value-balancing.com/_Resources/Persistent/2/6/e/6/26e 6d344f3bfa26825244ccfa4a9743f8299e7cf/20210210_VBA%20Impact%20S tatement_GeneralPaper.pdf. Accessed 12 December 2021.

VBA. 2021c. Methodology—Impact Statement: Extended Input-Output Modelling. Value Balancing Alliance. https://www.value-balancing.com/_ Resources/Persistent/0/f/9/1/0f919b194b89a59d3f71bd820da357804 5792e2c/20210526_VBA%20Impact%20Statement_InputOutput%20Mode lling.pdf. Accessed 12 December 2021.

WRI, WBCSD. 2004. A Corporate Accounting and Reporting Standard. World Resources Institute and World Business Council for Sustainable Development. https://ghgprotocol.org/sites/default/files/standards/ghg-protocol-revised.pdf. Accessed 02 February 2021.

WRI, WBCSD. 2011. Corporate Value Chain (Scope 3) Accounting and Reporting Standard. World Resources Institute and World Business Council for Sustainable Development. https://ghgprotocol.org/sites/default/files/ standards/Corporate-Value-Chain-Accounting-Reporing-Standard_041613_ 2.pdf. Accessed 02 February 2021.

CHAPTER 4

The Space Value of Money

Abstract This chapter presents the space value of money framework. It necessarily begins by introducing the analytical dimension of space, our physical context, into our heretofore risk and time focused analytical framework in finance. By doing so, it also introduces planet and humanity as two key stakeholders that must be formally accounted for in our mathematics of value and return alongside the mortal risk-averse return-maximising investor. The space value of money principle, which defines our relationship with space, establishes our spatial responsibility for impact. It offers a new space-adjusted mathematics of value and return with equations that quantify and integrate the space impact of cash flows into our models. The space value framework takes us beyond conventional risk and time based financial analysis and transforms the popular discounting methodology—the present value of future expected cash flows must now account for the space impact it would take to achieve or expect them, which must be compounded into the future when relevant.

Keywords Sustainability · Financial mathematics · Money · Value · Risk · Time · Space · Impact

JEL Classification E00 · E58 · G00 · G30 · Q51

© The Author(s), under exclusive license to Springer Nature Switzerland AG 2023
A. V. Papazian, *Hardwiring Sustainability into Financial Mathematics,*
https://doi.org/10.1007/978-3-031-45689-3_4

The analytical framework of the finance discipline reveals principles of value and equations where space, as an analytical dimension and our physical context, is missing. Indeed, as the discussion in Chapter 2 showed, our equations do not consider the impact of cash flows on and in space as an integral element of the value of cash flows. Furthermore, built around risk and time, focused on the non-actual future expected cash flows, omitting the actual space impact of cash flows, our equations serve the mortal risk-averse return-maximising investor and her/his/their preferences. Indeed, even sustainability is often treated as a taste or preference of the investor.

Chapter 3 reviewed the most prominent frameworks, standards, and tools aimed at operationalising sustainability in finance and revealed that they do not go far enough to penetrate the analytical content of core finance theory. Moreover, they fall short of transforming our core equations of value and return. Furthermore, the same risk bias can be found in our sustainability frameworks, where the consideration and integration of sustainability/ESG factors leads to adjustments to variables in our existing models.

This chapter presents the space value framework, which I have first introduced in Papazian (2022). The framework, naturally and necessarily, begins by introducing the analytical dimension of space, our physical context, into our analytical framework. By doing so, it establishes planet and humanity as the two key stakeholders that must be formally accounted for in our mathematics of value and return.

Following the introduction of the dimension of space, a new principle of value, the space value of money, becomes necessary. The principle defines our relationship with space, just like risk and return and time value of money define our relationship with risk and time. It complements the existing principles of value in finance and establishes the bottom threshold of investment acceptability.

A space-adjusted financial mathematics of value and return considers the space impact of cash flows to be an integral part of the value of cash flows. It offers a number of conceptual tools and a series of equations that quantify the space impact of cash flows and allow their integration into our models.

The proposed framework hardwires responsibility and sustainability into our equations. The value of future expected cash flows is no more dependent on their discounted present value alone, but also on their space impact in the present, which must be compounded into the future when relevant.

4.1   SPACE

Space is the dimension of 'where' and refers to our physical context of matter that stretches from subatomic particles to atoms and molecules within matter, to the planet, its core, surface, atmosphere, and outer space. Indeed, *space is defined as our physical context from subatomic to interstellar space and every layer in between and beyond.* Figure 4.1 provides a broad overview of the layers of space we affect and have interacted with (Smolin 2006; Papazian 2022, 130).

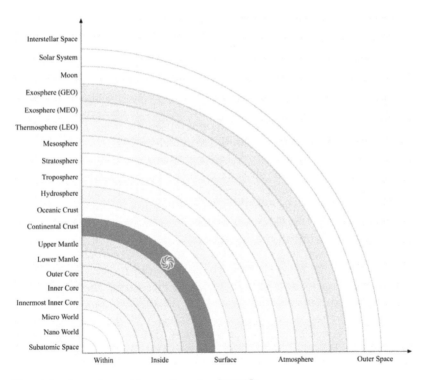

Fig. 4.1   Space layers (*Source* Papazian [2022])

It is important to note that every space layer mentioned in Fig. 4.1 can be further divided into sublayers. As in Table 4.1, we can see how the continental crust and hydrosphere (which includes the cryosphere) can be further broken down into sublayers as and when they are relevant to consider. The table provides further details on 'lakes' and 'vegetation' revealing that a fine-tuned understanding of each sublayer is also possible and should be considered when relevant.

The space layers identified in Fig. 4.1 and Table 4.1 present one plausible way of conceptualisation that reveals the diversity of contexts that our investments operate in and/or affect. The relevance of a space layer to a cash flow analysis is determined by the value chain of the investment under consideration.

However, we can also conceptualise the analytical dimension of space on a more abstract level, i.e., a three-dimensional space of length l. Indeed, a purely measurement-based conceptualisation of space layers is possible starting from a 3D space of length $= 10^0$ m (1 m) stretching towards the astronomical 10^{30} and the infinitesimal 10^{-30} ($l=10^y$).

While there is no outer boundary to a three-dimensional measurement of space, despite the limits of our own observable universe, our current understanding of matter reveals that when looking within, at the scale of the Planck length $= 1.616255 \times 10^{-35}$ m, our conventional understandings of space–time cease to apply and length stops making sense (Padmanabhan 1985; Seiberg 2006; Hossenfelder 2012).

I propose Fig. 4.2 as an alternative representation of space layers that could be used to depict the space layers involved in an investment. Naturally, using such an abstract conceptualisation of space layers does not automatically identify the specific nature or location of the layer. Thus, a combination of the two (Figs. 4.1 and 4.2) can be used to map the space impact of an investment and its value chain. Naturally, the same chart can be drawn at smaller scales, when focusing on a narrower spectrum. Figure 4.2 stretches from 10^{-24} to 10^{24} for convenience.

It is important to note that these layers, given their abstraction, can be defined from any point of matter on Earth, including the human body. While the space layers defined in Fig. 4.1 still apply, this provides a measurable conceptualisation of the layers. Table 4.2 provides the SI prefixes for each level of metric length in order to provide a more detailed understanding of each layer of space. The table also provides random examples of 'things' that could be found within each layer of space.

Table 4.1 Space layers' further details: hydrosphere and continental crust

Space layers	Sublayers	Sublayer type examples
Hydrosphere	*Seas, Lakes, Rivers, Ice Sheets*	Tectonic lakes
		Volcanic lakes
	Oceans	Glacial lakes
	Epipelagic Zone—The Sunlight Zone	Fluvial lakes
	Mesopelagic Zone—The Twilight Zone	Solution lakes
	Bathypelagic Zone—The Midnight Zone	Landslide lakes
	Abyssopelagic Zone—The Abyss	Aeolian lakes
	Hadal Zone—The Trenches	Shoreline lakes
		Organic lakes
		Anthropogenic lakes
		Meteorite lakes
Continental crust	*Land Surface*	Tundra
	Mountains	Taiga
	Built Up	Temperate broadleaf and mixed forest
		Temperate steppe
		Subtropical moist forest
	Vegetation	Mediterranean vegetation
		Tropical and subtropical moist forests
	Cropland	Arid desert
	Soil	Xeric shrubland
	O Horizon—Organic Layer	Dry steppe
	A Horizon—Top Soil Nutrient Layer	Semiarid desert
	E Horizon—Eluviation Layer	Grass savanna
	B Horizon—Subsoil Mineral Layer	Tree savanna
	C Horizon—Regolith Layer	Tropical and subtropical dry forest
	R Horizon—Bedrock Layer	Tropical rainforest
	Deep Crust	Alpine tundra
		Montane forest

Source Papazian (2022)

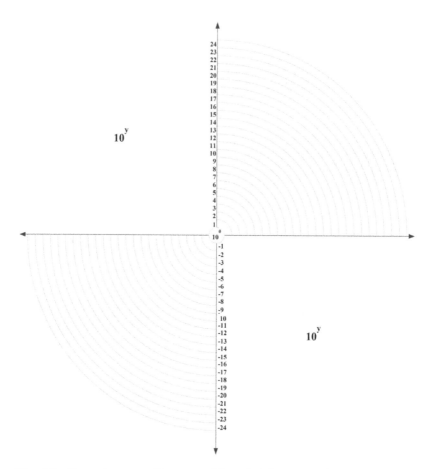

Fig. 4.2 Space layers as 3D space of length l in metres from any point of matter (l=10^y) (*Source* Author)

Having conceptualised the analytical dimension of space, our physical context, our next step is to introduce it into our analytical framework in finance. This requires that we also introduce *the space value of money principle*, a principle to define our relationship with space, just like the existing principles of finance define our relationship with risk and time.

Table 4.2 Space layers by length of 3D cube

Υ axis in metres	Equivalent	Random example in range—Diverse online sources
10^{30}	1 quettametre	New Prefixes adopted in November 2022 by the CGPM (2022)
10^{29}	100 ronnametre	
10^{28}	10 ronnametre	
10^{27}	1 ronnametre	
10^{26}	100 yottametres	The radius of the observable universe from Earth is approximately 435.19 yottametres
10^{25}	10 yottametres	GN-Z11 the high-redshift galaxy is 302.74 yottametres from Earth
10^{24}	1 yottametre	The distance to the Shapley Supercluster of galaxies is approximately 6.1 yottametres
10^{23}	100 zettametres	The Messier 87 supergiant elliptical galaxy is 501.41 zettametres from Earth
10^{22}	10 zettametres	The Andromeda galaxy is approximately 23.7 zettametres away from Earth
10^{21}	1 zettametre	The Milky Way galaxy is approximately 1 zettametre across
10^{20}	100 exametres	The cluster of young stars 'Westerlund 2' is approximately 189.21 exametres from Earth
10^{19}	10 exametres	The exoplanet Kepler-443b is approximately 28.38 exametres away from Earth
10^{18}	1 exametre	Our sun's closest twin star HIP 56948 is approximately 1.96 exametres away from Earth
10^{17}	100 petametres	The radius of the radio bubble emitted from Earth is approximately 946.073 petametres
10^{16}	10 petametres	The Alpha Centauri triple star system is 41.32 petametres from Earth
10^{15}	1 petametre	Light travels 9.461 petametres in one year
10^{14}	100 terametres	Light travels 777.062 terametres in one month
10^{13}	10 terametres	Voyager 1 was 23.953 terametres from Earth by 28/07/23, most distant human made object

(continued)

Table 4.2   (continued)

Υ axis in metres	Equivalent	Random example in range—Diverse online sources
10^{12}	1 terametre	Uranus is approximately 2.72395 terametres away from Earth
10^{11}	100 gigametres	Our sun is approximately 149.6 gigametres away from Earth
10^{10}	10 gigametres	Venus is approximately 41.4 gigametres from Earth
10^9	1 gigametres	Mars is approximately 7.834 gigametres away from Earth
10^8	100 megametres	The Moon is 384.399 megametres away from Earth
10^7	10 megametres	The Galileo Satellites orbit the Earth at an altitude of 23.222 megametres
10^6	1 megametres	The Cluster C1 satellite is on orbit at an altitude of 9 megametres
10^5	100 kilometres	The International Space Station (ISS) orbits the Earth at $\approx$402.336 kilometres from sea level
10^4	10 kilometres	Commercial airlines fly their airplanes at an average altitude of 10.6 kilometres
10^3	1 kilometre	Everest's summit is at 8.849 kilometres from sea level
10^2	1 hectometre	The height of the Empire State Building to its tip is 4.43 hectometres
10^1	1 decametre	The length of a wind turbine blade ranges between 5.2 and 10.7 decametres
10^0	1 metre	The average height of humans is 1.7 metres
10^{-1}	1 decimetre	The average length/hight of a full-term newborn human baby is 5 decimetres
10^{-2}	1 centimetre	A 2litre soda plastic bottle is 30 centimetres long
10^{-3}	1 millimetre	The average length of an adult mosquito is 4.5 millimetres
10^{-4}	100 micrometres	The average thickness of an eggshell is 300 micrometres or microns
10^{-5}	10 micrometres	The average thickness of a human hair is 70 micrometres or microns
10^{-6}	1 micrometre	The average thickness of a human red blood cell is 6–8 micrometres or microns

(continued)

Table 4.2 (continued)

Y axis in metres	Equivalent	Random example in range—Diverse online sources
10^{-7}	100 nanometres	The diameter of the Corona virus (SARS-CoV-2) ranges between 50 and 140 nanometres
10^{-8}	10 nanometres	The diameter of smallest viruses like Adeno-Associated Virus (AAV) is 20 nanometres
10^{-9}	1 nanometre	The diameter of a strand of human DNA is 2.5 nanometres
10^{-10}	100 picometres	The average size of a water molecule is 280 picometres
10^{-11}	10 picometres	The Bohr radius of a hydrogen atom is approximately 53 picometres
10^{-12}	1 picometres	The Compton wavelength of an electron is 2.4263 picometres
10^{-13}	100 femtometre	The diameter of the atomic nucleus of Uranium is approximately 11.7 femtometres or fermi
10^{-14}	10 femtometre	The radius of a gold nucleus is approximately 8.45 femtometres or fermi
10^{-15}	1 femtometre	The classical radius of an electron is 2.81 femtometres or fermi
10^{-16}	100 attometres	The approximate radius of a Proton is 841.8 attometres
10^{-17}	10 attometres	The range of the weak nuclear force is estimated to be 10 attometres
10^{-18}	1 attometre	The upper limit of the diameter of quarks which make up protons and neutrons in an atom
10^{-19}	100 zeptometres	
10^{-20}	10 zeptometres	Too small to give any non-technical
10^{-21}	1 zeptometres	examples
10^{-22}	100 yoctometre	
10^{-23}	10 yoctometre	
10^{-24}	1 yoctometre	
10^{-25}	100 rontometre	New Prefixes adopted in November 2022
10^{-26}	10 rontometre	by the CGPM (2022)
10^{-27}	1 rontometre	
10^{-28}	100 quectometre	
10^{-29}	10 quectometre	
10^{-30}	1 quectometre	

Source Author, inspired by Eames and Eames (1977) using NPL (2023)

4.2   The Space Value of Money

We have a missing principle in finance, a principle that establishes the value of cash flows vis-à-vis space—our physical context stretching from subatomic to interstellar space and every layer in between and beyond. I have proposed the introduction of a third principle into core finance theory and practice, the space value of money.

> The space value of money principle complements time value of money and risk and return. It establishes our spatial responsibility and requires that a dollar ($1) invested in space has at the very least a dollar's ($1) worth of positive impact on space. (Papazian 2022, 104)

The space value of money (SVoM) is the first step in transforming our value framework, and it leads to new equations of value and return where the responsibility of impact is integral to our models and equations. Indeed, while risk and return and time value of money define our relationship with risk and time from the perspective of the mortal risk-averse return-maximising investor, the space value of money introduces planet and humanity as equal stakeholders into our value framework.

The space value of money establishes the bottom threshold of investment acceptability because it requires that any dollar invested in space has at the very least a dollar's worth of positive impact on space, taking into account all the layers of space affected by the investment. To demonstrate the theoretical and practical relevance of the principle, consider the Transition Return Impact Map (TRIM) in Fig. 4.3. It denotes Space Impact and Investor Return and helps us visualise the transition challenge.

The challenge of the transition is to ensure that cash flows, investments, assets, projects, and companies do not have a negative space impact—whether we are considering emissions, other types of pollution and waste, or biodiversity loss. Our current financial value framework, given the focus on risk and time, does not *require* investors to have a positive space impact. Indeed, our current challenges and the many environments we have come to litter reveal that our framework and equations have to date absolved investors of their negative impacts thanks to the omission of space.

In other words, our current value paradigm in finance does not prevent investors from investing in opportunities or projects that are in the top left quadrant in Fig. 4.3 (Quadrant 3), where returns are positive, but impact

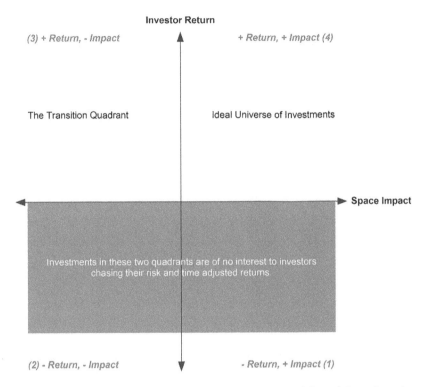

Fig. 4.3   TRIM: Transition return impact map (*Source* Adapted from Papazian [2022])

is negative. This is the transition quadrant, and it exists because we do not yet have a principle that prevents the undertaking of investments with negative space impact. Moreover, given our current value framework, the bottom two quadrants in Fig. 4.3 (Quadrants 1 and 2) are automatically dismissed by investors as unattractive—expected returns being negative, there is no reason to invest.[1]

---

[1] Note that actual returns could be negative post investment, and this is referring to expected and required returns. Also, some public investors may initiate investments in the bottom two quadrants for a variety of reasons, amongst them is the provision of public goods, paid through tax revenue.

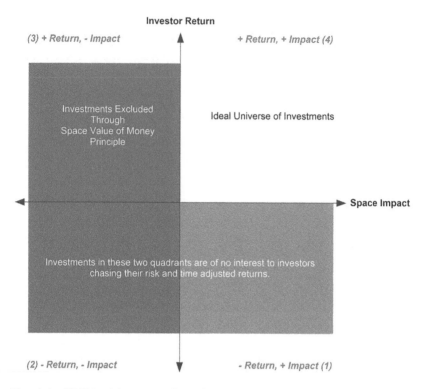

Fig. 4.4 TRIM with space value of money (*Source* Adapted from Papazian [2022])

The introduction of the space value of money (Fig. 4.4), therefore, acts as a bottom threshold of investment acceptability, as it excludes investments in projects, assets, and instruments that have a negative space impact. By doing so, the space value of money facilitates the introduction of new equations where humanity and planet are formally considered in our mathematics of value and return and the mortal risk-averse investor is no more the sole stakeholder of our models. The space value of money principle hardwires sustainability and allows us to, at the very least, prevent the inception of new investments that have a negative space impact.

The space value of money considers far more than just E, S, and G (environmental, social, and governance factors). It includes T and M as well (technology and money) and offers a systematic method that allows a thorough and authentic measurement of space impact across the many layers of space an investment may be affecting. The space value of money could be considered the theoretical link between sustainability and finance, and the principle that makes finance inherently sustainable.

4.3   A Financial Mathematics of Space Impact

Once we have introduced space as an analytical dimension, established our responsibility for space impact, and complemented our current principles with the space value of money, the next step is to devise a systematic method through which we can map and quantify space impact. To do so, we can start by expanding a commonly used tool in finance, the cash flow timeline. I have introduced the double timeline to depict the risk and space timelines that allow us to assess the space impact/value of cash flows along with their risk and time value (Fig. 4.5).

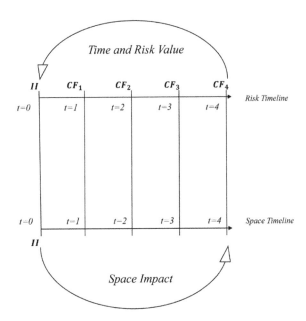

Fig. 4.5 Double timeline (*Source* Papazian [2022])

Given our new framework and the new principle, when we discount future expected cash flows to assess their time and risk value, we must also measure and compound the space impact of the cash flows across all affected space layers. The space impact it would take to achieve or expect the future cash flows must now be considered an integral part of the opportunity.

The minimum space value condition is that a dollar invested in space has at the very least a dollar's worth of positive impact on space. The below concepts and equations provide a conceptual summary of space impact using the Gross Space Value and Net Space Value equations, which measure the aggregate space impact of cash flows (Papazian 2022).

The minimum space value of money condition is met when Net Space Value (NSV) is equal to zero, and thus, the Gross Space Value is equal to the initial investment—when a dollar invested has a dollar's worth of positive space impact.

$$\text{Gross Space Value}_{T,S} = \text{NSV} + \text{II} \qquad (4.1)$$

$$\text{NSV} + \text{II} = \text{GSV} \mid \text{NSV} = 0$$

$$\text{GSV} = \text{II}$$

where the Net Space Value is broadly defined as the planetary, human, and economic impact of the investment or cash flows—summarised as:

$\text{NSV}_{T\&S} = \text{Net Space Value of Investment}$

$\text{NSV}_{T\&S} = \{\text{Planetary, Human,}$

$\qquad \text{and Economic Impact}\}_{\text{All } Space \text{ Layers \& } Time \text{ Periods}}$

$T = \text{Total Number of Years of Investment being Considered}$
$S = \text{All Space layers Involved in the Investment}$

$$NSV_{T\&S} = \sum_{t=1}^{T} \sum_{s=1}^{S} \text{Pollution \& Biodiversity Impact}$$

$$+ \sum_{t=1}^{T} \sum_{s=1}^{S} \text{Human Capital \& R and D Impact}$$

$$+ \sum_{t=1}^{T} \sum_{s=1}^{S} \text{New Asset \& New Money Impact} \qquad (4.2)$$

where each of the elements, planetary, human, and economic are further defined as the pollution, biodiversity, human capital, R & D, new money, and new asset impacts of the investment (see Table 4.3).

The next subsections discuss the component equations of Net Space Value in more detail. A few general observations are due here. Pollution and biodiversity elements consider the clean-up costs (C_{pst}) and restoration costs (R_{bst}) in order to quantify the projected costs of the cash flows and investments across all time period, space layers, and types of pollutants and biodiversity losses, and they do so for prevention purposes; they are not to be confused with direct and/or indirect encouragement of post event treatment.

Given the space value principle, negative impacts must be identified across all categories and prevented, and positive impacts do not absolve the investment from its negative impacts—offsetting is not considered to be a viable way of meeting the space value of money principle. Furthermore, across all these component equations, the main purpose is the development of a comprehensive aggregation tool, as such, impact assessment tools like the Life Cycle Approach can still be used.

All the coefficients, i.e., *fairness, health, governance, and transition,* naturally involve a certain level of subjective assessment—in other words, the framework does not pretend to be 'subjectivity free.' However, unlike ESG ratings and other proposed metrics like Implied Temperature Rise (ITR) rates, the space value equations offer a framework through which marginal new information can be interpreted by the market, even if traders may sometimes reach contradictory conclusions given their subjective assessments. This is a key aspect missing in the ESG ratings-based approach, where the absence of equations and a replicable framework makes it impossible for the market to interpret, and potentially act upon,

Table 4.3 Equations of impact

Impact aspect	Net space value	
	$g \times (\mathbf{PI}_{T,S,P} + \mathbf{BI}_{T,S,B} + \mathbf{HCI}_{T,S} + \mathbf{RDI}_{T,S,N} + \mathbf{NAI}_{D,S,A} + \mathbf{NMI}_T)$	(4.3)
Planetary	Pollution Impact $\quad PI_{T,S,P} = \sum_{t=1}^{T}\sum_{s=1}^{S}\sum_{p=1}^{P} Q_{pst} \times C_{pst}$	(4.4)
	Biodiversity Impact $\quad BI_{T,S,B} = \sum_{t=1}^{T}\sum_{s=1}^{S}\sum_{b=1}^{B} A_{bst} \times R_{bst}$	(4.5)
Human	Human Capital Impact $\quad HCI_{T,S} = f \times \sum_{t=1}^{T}\sum_{s=1}^{S} E_{st} + T_{st} + H_{st} + I_{st} + C_{st} + S_{st}$	(4.6)
	R and D Impact $\quad RDI_{T,S,N} = \sum_{t=1}^{T}\sum_{s=1}^{S}\sum_{n=1}^{N} h_n \times RD_{tsn}$	(4.7)
Economic	New asset Impact $\quad NAI_{D,S,A} = \sum_{s=1}^{S}\sum_{a=1}^{A} k_a \times BVA_{asD}$	(4.8)
Coefficients	New Money Impact $\quad NMI_T = (II \times DR \times BLR) + mm \times (II + X_T - M_T)$	(4.9)
	Fairness $\quad f$	
	Health $\quad b$	
	Transition $\quad k$	
	Governance $\quad g$	

Source Adapted from Papazian (2022)

a new piece of marginal information that is expected to affect a specific rating.

The space value framework introduces the necessity and tools to assess the impact of cash flows across *all the layers of space* that the considered cash flows may be affecting. This is a unique and an important distinction for a variety of reasons:

$$\sum_{s=1}^{S}$$

a. Value Chain: because the value chains of investments affect different layers of space—a shipping company that uses aeroplanes affects the stratosphere, while a shipping company using ships affects the hydrosphere.
b. Impact Intensity: because the intensity of impact differs across space layers, GHG emissions in the stratosphere have a different impact from GHG emissions in busy densely populated cities with high-rise buildings.
c. Clean-up Technology: because cleaning the same pollutant in different space layers involves different technologies—cleaning plastic from our oceans, from our rivers, from our streets, and from our food chain require different technologies.
d. Costs: given all of the above, the costs of impact differ across different layers of space, even for the same pollutant or type of waste.

Given the detailed discussion and presentation of these equations in Papazian (2022), the analysis below is focused on introducing the equations.

4.3.1   *Planetary Impact: Pollution and Biodiversity Impacts*

4.3.1.1   *Pollution and Waste*
The first aspect of planetary impact concerns pollutions of all kinds, including waste. The encyclopaedia Britannica defines pollution as follows:

> [T]he addition of any substance (solid, liquid, or gas) or any form of energy (such as heat, sound, or radioactivity) to the environment at

a rate faster than it can be dispersed, diluted, decomposed, recycled, or stored in some harmless form. The major kinds of pollution, usually classified by environment, are air pollution, water pollution, and land pollution. Modern society is also concerned about specific types of pollutants, such as noise pollution, light pollution, and plastic pollution. Pollution of all kinds can have negative effects on the environment and wildlife and often impacts human health and well-being. (Britannica 2023)

Based upon the above, I define pollutants to be those materials, substances, elements, particles, nano-particles, and molecules that are known to harm the environment and human and animal well-being and are generated through the design, testing, industrial or non-industrial production, transportation, distribution, consumption, and disposal of final and intermediary products and services across all sectors of the economy.

The pollution impact of an investment across all space layers (S) and over the lifetime of the investment (T) can be quantified *by multiplying the pollutants that are being created or left behind multiplied by the cost of removal or clean-up.*

These costs can be calculated either directly or through proxies like recycling costs (Trucost 2013; WRAP 2018). When looking at carbon or emissions, the clean-up cost should not be confused with Carbon Prices (Ellerman et al. 2010) or the Effective Carbon Rate (OECD 2018, 2021), although such mechanisms could be used when relevant. GHG emissions and conversion rates could be used to standardise GHG quantities and costs (HM Government 2019, 2021). The measurement of the pollution impact of investments can also benefit from the literature on negative environmental externalities—for example, the Life Cycle Assessment approach can be used to plot and identify such impacts for products and services (SETAC 1991, 1993; EIA 1995; Nguyen et al. 2016).

$$\text{Pollution Impact}_{\text{All Pollutants Across Time and Space}} = PI_{T,S,P} =$$

$PI_{T,S,P}$ = Sum of All Individual Pollutant Impacts Across Years and Space Layers

$$PI_{T,S,P} = \sum_{t=1}^{T} \sum_{s=1}^{S} \sum_{p=1}^{P} Q_{pst} \times C_{pst} \qquad (4.10)$$

T = Total Number of Years in the Investment
P = Number/Types of Pollutants Involved
S = All Space Layers Involved
Q_{pst} = Quantity of Pollutant p in time t in space layer s
C_{pst} = Cost of Cleanup for Pollutant p in time t in space layer s

The above equation measures the total pollution impact of an investment across all the time periods of the investment and across all the space layers affected by the investment, and it considers all pollutants created across the entire value chain of the investment. This is not about carbon, or plastic alone, and can cover nano-toxins and space debris as well if they are being created due to the investment under consideration.

We can also adjust the above equation to consider specifically the impact of one particular pollutant (Individual Pollutant Impact—IPI). This must take into account the fact that the quantity and cost of clean-up may differ across the space layers and time periods involved in the investment. This is so because cleaning a specific pollutant from land surfaces may be cheaper than cleaning or removing the same pollutant from outer space.

$$IPI_{T,S,P_1} = \sum_{t=1}^{T} \sum_{s=1}^{S} Q_{p_1st} \times C_{p_1st} \qquad (4.11)$$

$IPI_{T,S,P}$ = Individual Pollutant Impact Across Years and Space Layers

If the investment is removing or absorbing a pollutant, so it is cleaning it up, the metric still applies, but as a positive value. In other words, the positive or negative sign attached to this measure depends on whether the investment is creating pollution or cleaning up pollution.

4.3.1.2   Biodiversity

The second most critical aspect of planetary impact is biodiversity impact (IPBES 2019). White et al. (2021) summarise the biodiversity challenge as follows:

> Despite increasing recognition of its importance, biodiversity is in precip-
> itous decline (Díaz et al., 2019; Tittensor et al., 2014). Recent reports
> estimate that 75% of the terrestrial environment and 66% of the marine
> environment have been severely altered by human activity (Halpern
> et al., 2015; IPBES, 2019; Venter et al., 2016), and that between 1970
> and 2014 populations of monitored species have declined by an average
> of 70% (WWF, 2018b). This decline is largely driven by the continued
> growth of the global economy (Hooke et al., 2012; IPBES, 2019; Maxwell
> et al., 2016). From aquaculture and forestry to mining, consumer goods,
> and infrastructure, industrial development across sectors is closely tied to
> biodiversity loss. Business operations and supply chains act to increase the
> production and movement of goods, often at the expense of natural ecosys-
> tems through increasing habitat loss, fragmentation, pollution, invasive
> species introductions, and overexploitation (Díaz et al., 2019; Krausmann
> et al., 2017). Consequently, biodiversity loss is recognized as a major global
> challenge for the private sector presenting operational, financial, and repu-
> tational risks (Global Canopy & Vivid Economics, 2020; WEF, 2021).
> (White et al. 2021)

Whatever the type of biodiversity in question, i.e., genetic, species, or ecosystem biodiversity, and whatever the type of threat involved, i.e., habitat loss and destruction, altered composition, introduction of exotic species, overexploitation, pollution and contamination, or climate change, and whatever the specific environments they affect, i.e., marine, land, air, and even one day other planets, and whatever the level of diversity within those habitats, for the purposes of an all-encompassing conceptual metric for biodiversity loss measurement in monetary terms, the concept of biodiversity impact must be measured at the most abstract level, i.e., the area or volume of space affected (ESA 2017; Papazian 2022, 139).

The metric I propose for the measurement of biodiversity impact uses the cost of restoration or habitat replacement costs (Environment Agency 2015) as an indicative measure of the biodiversity cost of the investment. The feasibility of habitat restoration (HM Government 1994; Parker 1995) is significant, and a similar concept of Habitat Restoration Cost

(HRC) is discussed at length in the natural capital literature (Shepherd et al. 1999; Pearce and Moran 1994).

Given the space value of money principle, the purpose here is to prevent negative biodiversity impact by integrating their projected restoration costs in the value of expected future cash flows.

$$\text{Biodiveristy Impact}_{\text{Across Time and Space}} = BI_{T,S,B} =$$

$$BI_{T,S,B} = \sum_{t=1}^{T} \sum_{s=1}^{S} \sum_{b=1}^{B} A_{bst} \times R_{bst} \qquad (4.12)$$

T = Total Number of Years in the Investment
S = All Space Layers Involved
B = All Types of Biodiversity Involved
$BI_{T,S}$ = Sum of All Biodiversity Impacts Across Years and Space Layers
A_{bst} = Area (ha) of Biodiversity Impact b in time t in space layer s
R_{bst} = Cost of Restoration $\left(\frac{\$}{\text{ha}}\right)$ of Biodiveristy Impact b in t in s

Similarly to the pollution metric used in the previous section, when an investment is restoring a habitat and biodiversity, the measure still applies, not as a loss of value, but as a gain. Once again, the positive or negative sign attached to this measure depends on whether the investment is causing or restoring biodiversity loss.

* * *

To conclude the planetary impact section, putting the pollution and biodiversity equations together, we get:

Planetary Impact = PLANETI = Pollution Impact + Biodiveristy Impact

$$PLANETI_{T,S} = PI_{T,S,P} + BI_{T,S,B}$$

$$PLANETI_{T,S} = \sum_{t=1}^{T} \sum_{s=1}^{S} \sum_{p=1}^{P} Q_{pst} \times C_{pst} + \sum_{t=1}^{T} \sum_{s=1}^{S} \sum_{b=1}^{B} A_{bst} \times R_{bst} \quad (4.13)$$

4.3.2   Human Impact: Human Capital and R&D Impacts

4.3.2.1   Human Capital

At the heart of all human productive activities lies human capital, a key and defining feature of any investment. While technology, hardware and/ or software, is a growing and central factor in human productivity, human capital remains an essential aspect of all investments and it is defined as:

> Human capital consists of the knowledge, skills, and health that people accumulate over their lives. People's health and education have undeniable intrinsic value, and human capital also enables people to realize their potential as productive members of society. More human capital is associated with higher earnings for people, higher income for countries, and stronger cohesion in societies. It is a central driver of sustainable growth and poverty reduction. (World Bank 2020, 1)

Indeed, however minimally or marginally, human capital is involved in almost all investments. Even a technology-driven stock market transaction in the secondary market, where the invested cash does not reach the traded corporation, involves human capital. While not immediately visible, such a transaction makes use and supports the value chains of the brokerage, the custodian bank, and the stock exchange through which the 'online trade' is executed and settled.

Thus, the first step in identifying the human capital impact of an investment is to identify the nature of the involvement. An investment in a productive value chain where human capital is employed implies a direct involvement of human capital, while in the above secondary market transaction example, the investment implies an indirect involvement of human capital after investment, due to the follow-up use of the fees/investment.

Not all investments involve human capital, and when they do, they do not all have the same type of deployment and use—some involve supporting investments, others do not, some are based on fair treatment, consideration, wages, and compensation, others not.

Current accounting standards do not consider employment as an asset. Employment is an expenditure, combining salaries, benefits, etc., which together can reveal the human capital impact of an investment. I propose the below equation that groups such expenditures into five subcategories, together ETHICS, representing Employment, Training, Health,

Immigration, Compensation, and Social Investment Expenditures.

$$\text{Human Capital Impact} = \text{HCI}_{T,S}$$

$\text{HCI}_{T,S}$ = Fair and responsible expenditure on humans within and around

$$\text{HCI}_{T,S} = f \times \sum_{t=1}^{T} \sum_{s=1}^{S} E_{st} + T_{st} + H_{st} + I_{st} + C_{st} + S_{st} \qquad (4.14)$$

S = All Space Layers Involved
T = Total Number of Years in the Investment
f = Coefficient of Fairness
E_{st} = Employment Expenditure
T_{st} = Training and Education Expenditure
H_{st} = Health Related Expenditure
I_{st} = Immigration Related Expenditure
C_{st} = Compensation Expenditure
S_{st} = Social Investment Expenditure

The coefficient f ($-1 \leq f \geq 1$) identifies the level of fairness in the organisation and/or in the employment of the human capital being used in the investment. This, amongst other features, can be defined by gender balance, equal pay, and fair treatment of employees. Naturally, the value assigned to f in a specific investment involves a subjective and objective analysis of the operations and contractual processes that govern the employment and deployment of human capital through the considered investment.

A fair and well-managed human capital expenditure that respects our requirements for gender balance, fair wages, equal pay, and fair treatment can receive a positive value, and an investment where the human capital process is neither fair nor equal can receive a value of zero, or a value less than 1.

4.3.2.2   Research and Development

All human technological advances are born from a process of innovation that is built and dependent upon research and development (R and D). Ideas and insights become tangible economic value through the efforts put into researching and developing them into actual products and services. This is so for all fields and subjects, including those that are metaphysical in nature. R and D are central to any process of transformation, and especially critical for a species faced with the challenges posed by climate change and ecological destruction.

A recent OECD report, the Frascati Manual, defines R and D expenditure as follows:

> Research and experimental development (R&D) comprise creative and systematic work undertaken in order to increase the stock of knowledge – including knowledge of humankind, culture and society – and to devise new applications of available knowledge.... The term R&D covers three types of activity: basic research, applied research and experimental development. (OECD 2015, 44)

Here, R and D expenditure is taken into account as a value creating activity that enhances knowledge, experience, and stretches the boundaries of knowledge and/or a productive or business process. OECD (2015) also identifies R and D projects as the conceptual unit that helps quantify a specific set of R and D activity or activities. This is so because investments and companies can always involve simultaneous R and D projects aimed at the development of entirely different products or technologies. This is the approach adopted in the below proposed equation that quantifies the R and D impact of an investment or cash flows.

An investment that includes R and D activity differs from an investment that does not, and the expenditures allocated to such activity and projects are the right measure to consider when assessing the R and D impact of such an investment. Such expenditure must naturally include current and fixed capital expenditures and must also take into account the fact that employment expenditures, as discussed in the previous section, are already considered in the human capital element of impact assessment.

$$R \text{ \& } D \text{ Impact} = RDI_{T,S,N}$$

$$\mathrm{RDI}_{T,S,N} = \sum_{t=1}^{T} \sum_{s=1}^{S} \sum_{n=1}^{N} h_n \times \mathrm{RD}_{tsn} \qquad (4.15)$$

h = Coefficient of health

$\mathrm{RD}_{T,S,N}$ = R & D Expenditure per Project N across T and space layers S

S = All Space Layers Involved

T = Total Number of Years in the Investment

N = Number/All R and D Projects Involved in the Investment

When mismanaged, an R and D process can damage the environment and human health. The example of Teflon and C8, or 'eternal chemicals' in general, demonstrates that R and D can be an unhealthy process and can be so badly managed that it leads to deaths and extensive environmental harm. In parallel, an R and D process could be unhealthy for those involved and their families depending on industry and practice. Thus, a coefficient of health is necessary when quantifying and assessing the impact of R and D expenditures.

The coefficient h_n, $(-1 \leq h_n \geq 1)$ is a qualitative assessment of the healthy management of the R and D projects being considered, taking into account safety measures included for humans and the environment.

* * *

Combining the two aspects of human impact, Human Capital and R and D Impact, we can write the total as follows:

Human Impact = HUMANI = Human Capital Impact + R & D Impact

$$\mathrm{HUMANI} = \mathrm{HCI}_{T,S} + \mathrm{RDI}_{T,S,R}$$

$$\mathrm{HUMANI} = f \times \sum_{t=1}^{T} \sum_{s=1}^{S} E_{st} + T_{st} + H_{st} + I_{st} + C_{st} + S_{st}$$
$$+ \sum_{t=1}^{T} \sum_{s=1}^{S} \sum_{n=1}^{N} h_n \times \mathrm{RD}_{tsn} \qquad (4.16)$$

4.3.3   Economic Impact: New Assets and New Money Impacts

The omission of space impact from our equations of value and return, as discussed in Chapter 2, revealed that conventional finance models and equations have ignored the dimension of space. This omission did not just involve pollution, biodiversity, human, and R and D impact on space, but also the economic impact of cash flows and investments.

Thus, the assessment of space impact from an economic standpoint, i.e., the new assets and new money that an investment creates, is also relevant to this discussion. Indeed, while some investments create new technology, new real estate, new inventory, etc., others do not. This implies that their space impact cannot be assumed to be identical, and the economic impact of cash flows and investments deserves the same level of attention.

4.3.3.1   New Assets

Investments and cash flows do not have an identical asset impact. Even when their financials are identical, and they are in the same sector and industry, the asset footprint of cash flows is defined by the mode of utilisation and deployment of the cash flows.

> For example, think of two investments, both requiring a $1 million dollar investment, both have the same payback period, both have the same risk, both have the same return to the investor, and both spend $300,000 on real estate. Do they have the same impact? Not really, because the first investment is using the $300,000 to rent its real estate in the city over the lifetime of the investment, and the second investment is buying a small plot of land outside the city and developing its own solar energy powered real estate. Similarly, two companies that are in the solar energy business, one is developing its own new technology, the other is a distribution outlet. Both happen to have the same return on investment, but they have very different asset impacts given the creation of new intellectual property and new technology. (Papazian 2022, 147–148)

The asset impact of an investment is a central feature of the space impact of cash flows. Thus, it is paramount to identify the new assets that the investment will create across all the layers of space it affects or operates in. To calculate the new assets created through an investment, we can apply the following formula, where we add the value of all tangible and intangible assets created across all space layers:

$$\text{New Asset Impact} = \text{NAI}_{D,S,A} = \sum_{s=1}^{S} \sum_{a=1}^{A} k_a \times \text{BVA}_{asD} \qquad (4.17)$$

k_a = Coefficient of Transition Value
S = All Space Layers Involved
D = Period/Date When Asset is Created and Added in Books
A = All Tangible and Intangible Asset Created through the Investment
BVA_{asD} = Book Value of Asset a in space layer s recorded at date D

The equation covers all types of assets, tangible and intangible, real and financial. While Eq. 4.17 considers the book value of the assets, in some specific cases, we can use the market value of assets as well, depending on the type and volatility of the created asset. Note that the assets are added once, as per the date (D) when they are added on the books.

The coefficient of transition, K_a, is defined on the individual asset level, and it is a coefficient with a value between $-1 \le k_a \ge 1$ and it can be used to qualify the assets being created vis-a vis the transition. This is of crucial importance given the global Net Zero agenda and the planned and expected transformations of the world economy in the near future. This ensures that we are not valuing a coal mine and a carbon capture technology plant in the same identical way. This integrates the 'stranded assets' proposition, "assets [that] suffer from unanticipated or premature write-offs, downward revaluations or are converted to liabilities" (Caldecott et al. 2013, 7). Indeed, we could assign k_a a value of zero or one, depending on the considered asset's relevance to the low carbon future we are unavoidably headed towards.

While the new asset impact of an investment is important and relevant to the assessment of the overall space impact of the considered investment, it does not absolve it from negative planetary and human impacts.

An investment may create new assets and may also cause the creation of new money in the economy. The monetary impact of an investment is also part of its space impact and should be properly accounted for.

4.3.3.2   New Money

Quantifying the monetary impact of an investment begins at the source. An investment can be a cash payment/deposit into an account or a wire transfer. As such, the cash in banknote form or digital form must already exist. In some cases, an investment is achieved through a bank loan or other forms of debt or leverage where the invested amount is itself new money.

The first step when assessing the monetary impact of an investment, therefore, must identify whether the invested money is new or existing money. An equity investment from a private equity firm involves money already in existence and involves the transfer of money from one account to another. A shareholder loan that supports the working capital of a firm is existing money being transferred to the company account. However, a bank loan (through central or commercial banks) involves the creation of new money through the banking system.[2]

Thus, the first step in quantifying the new money impact of any investment is to identify which proportion, if any, of the investment amount is new money. We achieve this by identifying the proportion of bank debt involved in the initial investment.

$$\text{New Money Impact} = \text{NMI}_T = \text{II} \times \text{DR} \times \text{BLR} \qquad (4.18)$$

II = Initial Investment
DR = Debt Ratio
BLR = Bank Loan Ratio
NMI_T = Total across the Years in the Investment

Whether the investment is new money or existing money, it starts off as a transfer of some kind, which can be treated as a new deposit somewhere. Given the debt-based nature of our monetary architecture, this new deposit may in turn lead to new money creation within the economy.

> If we were to treat the initial investment as a new deposit, whether partly or entirely new money, or entirely existing money, we could also measure its impact on the money creation cycle as it is spent within the economy,

[2] I discuss central bank and commercial bank new money creation in detail in the next chapter.

becoming new deposits in other banks. To measure how much money creation potential the investment implies as it is spent in the economy, if it is, we can multiply it with the *actual* money multiplier.[3] This is proposed knowing full well the limitations of the money multiplier concept. In this equation the money multiplier is the actual multiplier as observed in the economy at time t−1, and it is used as a descriptive measure. (Papazian 2022, 151)

Thus, we could also expand equation (Eq. 4.18) to include the money creation potential of the investment as it contributes to new deposit creation in the banking system. To reflect a more authentic picture, in truth, we should also consider planned imports and expected exports as well. If the investment involves importing raw materials, then some of the investment will leave the economy through foreign exchange. Similarly, if the investment aims to export goods and services, the business is earning money through foreign exchange.

To adjust and add the money creation potential of the investment, we add expected exports and subtract planned imports from the initial investment, and then multiply the sum $(II + X - M)$ by the money multiplier.

Given that the monetary impact, as of today, occurs in one space layer, and given that it has to be considered across the entire investment time window, it is written as follows:

$$\text{New Money Impact} = \text{NMI}_T$$

$$\text{NMI}_T = II \times DR \times BLR + mm \times (II + X_T - M_T) \qquad (4.19)$$

mm = Money Multiplier
DR = Debt Ratio
BLR = Bank Loan Ratio
M_T = Planned Imports
X_T = Expected Exports

[3] I use the actual money multiplier as observed in $t - 1$, as a descriptive measure. Studies have argued about the irrelevance of the money multiplier for central bank policy transmission, also clarifying the relationship between the money multiplier and bank credit policy. Thus, this is only used as an after the fact observation not policy tool (McLeay et al. 2014; Ihrig et al. 2021).

II = Initial Investment
NMI_T = Total across the Years in the Investment

The new money created through an investment is particularly relevant for large and public investments. However significant, a high positive new money impact does not absolve the investment of its responsibilities for negative planetary and human space impact.

* * *

To put together the economic impact in one integrated equation:

ECONOMIC Impact = ECONOMICI = New Asset Impact + New Money Impact

$$ECONOMICI = NAI_{D,S,A} + NMI_T$$

$$ECONOMICI = \sum_{s=1}^{S} \sum_{a=1}^{A} k_a \times BVA_{asD}$$
$$+ [II \times DR \times BLR + mm \times (II + X_T - M_T)] \qquad (4.20)$$

4.3.4   Governance

Governance has been centre stage in sustainable finance, and ESG ratings and factor integration have played an important role in operationalising the role of governance in the field. However, good governance has been a key subject of research for many decades, and it has been studied in relevance to the performance of governments and government institutions, as well as private and public corporations.

Corporate governance has been a central topic in finance and management research. Managerial incentives, management turnover, board composition, succession, committees, contracts, and shareholding structures are amongst the many aspects that have been studied at length, looking at good governance and agency costs associated with corporations and the pursuit of shareholder value (Jensen and Meckling 1976; Fama 1980; Fama and Jensen 1983; Shleifer and Vishny 1989, 2012; Gilson 1989; Bluedorn 1982; Brickley and Drunen 1990; Dissanaike and Papazian 2005; Papazian 2004; Bhagat et al. 2008; Daines et al. 2010).

[Indeed], governance plays a key role in the way an investment is managed, which affects both the *outcomes* and *impacts* of the investment. In other words, governance affects both, the expected future cash flows, and the space impact of the investment.... [f]rom an impact assessment perspective, the key purpose of the governance factor is to reveal or assess the reliability and veracity of the facts and figures being reported - facts and figures upon which the entire analysis of impact is based. (Papazian 2022, 153)

As such, I have proposed the coefficient g to denote the investor's degree of belief in the truthfulness of the reported figures and facts, and the degree of trust in the authenticity of commitment to targets and objectives. This is a summary coefficient that assesses the assumed level of good governance.

$$g = \text{Coefficient of Good Governance}$$

The coefficient g may or may not be relevant to a specific investment. When it is, it should be a coefficient between $0 \leq g \leq 1$ reflecting the level of trust in the veracity of the figures based on the corporate governance policies and structures in the investment. The coefficient can be used to adjust the Net Space Value of Money equation as follows:

$$\text{Net Space Value}_{T,S} = g \times \big(\text{PI}_{T,S,P} + \text{BI}_{T,S,B} + \text{HCI}_{T,S}$$
$$+ \text{RDI}_{T,S,N} + \text{NAI}_{D,S,A} + \text{NMI}_T\big)$$

When governance is considered to be a factor, the next key step is to identify if the exposure is to a single entity or multiple entities. In a primary market equity investment, investors are exposed to the governance structure of the business they are investing in. However, if they invest through funds and/or asset management firms, or through a syndicated investment program, they are exposed to the governance processes of more than one entity. In such cases, each could be qualified with a different g_1 and g_2 coefficient.

When governance is a factor, whether the exposure is to a single entity or multiple entities, the main purpose of the coefficient g is to establish the degree of trust in the facts and figures and plans involved in the investment.[4]

---

[4] It is important to note here that the commonly assessed factors under G in ESG are treated and considered in the human capital aspect of space impact measurement.

4.3.5   Impact Intensities

Given the many aspects and elements of space impact included in a Net Space Value calculation, as discussed in the previous sections, the space value framework allows the introduction of three intensity measures which provide us the opportunity of a more granular assessment of impact, going far deeper than carbon intensity.

The planetary, human, and economic impact intensities of investments and cash flows introduce the opportunity to identify the specific footprint involved. Naturally, these intensity measures can be calculated for past actual and/or future expected cash flows.

Planetary Impact Intensity$_t$ = PI and BI per Expected Cash Flow$_t$

$$\text{PLANETII}_t = \frac{\text{PI}_{t,S,P} + \text{BI}_{t,S,B}}{\text{Expected Cash Flow}_t} \qquad (4.21)$$

Human Impact Intensity$_t$ = HCI and RDI per Expected Cash Flow$_t$

$$\text{HUMANII}_t = \frac{\text{HCI}_{t,S} + \text{RDI}_{t,S,N}}{\text{Expected Cash Flow}_t} \qquad (4.22)$$

Economic Impact Intensity$_t$ = NAI and NMI per Expected Cash Flow$_t$

$$\text{ECONOMICII}_t = \frac{\text{NAI}_{d,S,A} + \text{NMI}_t}{\text{Expected Cash Flow}_t} \qquad (4.23)$$

4.3.6   The Space Growth Rate

Another key contribution of the space value framework is the space growth rate (SPR), which measures the implied periodic rate of growth that takes us from the initial investment (II) to the aggregate Net Space Value of the investment across the T periods of the investment. The space growth rate is a summary rate that, in a way, mirrors the discount rate.

As depicted in Fig. 4.6, and given the space value of money principle, while investors can continue pursuing their risk and time-adjusted returns, their impact must be accounted for and compounded into the future when relevant. Equations 4.24 and 4.25 reveal the relationship between the space growth rate and the Net Space Value of an investment, where the investment is considered as a series of cash expenditures, and the SPR is used to compound them into the future. In truth, just like benchmark discount rates being used in markets, we could also have benchmark space growth rates >0 that would set the minimum required positive space impact for public and private investments.

$$\text{SPR} = \sqrt[T]{\frac{\text{NSV}_{T,S}}{\text{II}}} - 1 \qquad (4.24)$$

SPR = The Space Growth Rate per period

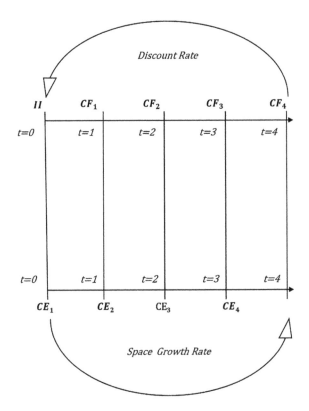

Fig. 4.6 Double timeline and the space growth rate (*Source* Papazian [2022])

II = Initial Investment
T = All Time Periods involved in the Investment
S = All Space Layers Involved in the Investment

The space growth rate allows us to conceptualise the Net Space Value as the compounded future value of the cash expenditures (CE_t) that make up an initial investment.

$$\text{Net Space Value} = NSV_{T,S}$$

$$NSV_{T,S} = \sum_{t=0}^{T} CE_t (1 + SPR)^{T-t} \tag{4.25}$$

4.3.7   Integrating Impact into Value

The previous sections introduced the analytical dimension of space, the principle of value that defines our relationship with space, the double timeline to help us integrate the risk, time, and space value of cash flows, the concepts of Gross and Net Space Value of investments, the planetary, human, and economic aspects of impact and their components, the many impact intensities that reflect the dependence of cash flows on these impacts, and the space growth rate.

The next step is to discuss how such measurements of space impact can be integrated into our value and return equations. In order to entrench sustainability into our value equations, investments and cash flows must be valued considering their space impact. This implies that when negative impacts are possible and part of the impact of an investment and its cash flows, then those negative impacts must be made to affect the value/return of the investment negatively. In the following sections, I discuss two examples of space impact/value integration.

4.3.7.1   Space Impact Adjusted Present Value: NPV

The space value of money principle requires that a dollar invested in space has at the very least a dollar's worth of positive impact on space. As discussed in the previous sections, the principle establishes the bottom threshold of investment acceptability by requiring the identification, quantification, and exclusion of negative impacts from the investment process, and thus the productive value chain of investments.

In other words, when an investment or expenditure that aims to generate future expected cash flows has a negative space impact, whether planetary, human, or economic, that negative impact must affect its value. This consideration or integration must ideally be done before the investment is taken on, for prevention purposes and to reflect the true cost of the investment—to reveal the true present value that takes into account the space impact of the cash flows in the present.

When considering the above in the case of the Net Present Value equation, therefore, we must consider the negative space impacts involved in the investment. Thus, the NPV equation must be adjusted by the negative space impacts of the initial investment if and when they occur.

$$\text{NNSV}_{T,S} = \text{The Sum of Negative Impacts Across All Years and Space Layers}$$

NNSV is the sum of all the negative space impacts involved in the investment across all space layers and time periods involved. It is *not* the negative of the total Net Space Value of the investment, but the sum of the negative elements within NSV. NNSV identifies the monetary value of the negative impacts that will take to achieve the future expected cash flows, on top of the initial investment.

$$\text{Net Space Value}_{T,S} = g \times \big(\text{PI}_{T,S,P} + \text{BI}_{T,S,B} + \text{HCI}_{T,S}$$
$$+ \text{RDI}_{T,S,N} + \text{NAI}_{D,S,A} + \text{NMI}_T\big)$$

We can thus rewrite the NPV equation, as the negative impact adjusted NPV as follows:

Negative Impact Adjusted Net Present Value = NIA NPV

$$\text{NIA NPV} = -\left|\text{NNSV}_{T,S}\right| - \text{II} + \sum_{t=1}^{T} \frac{\text{CF}_t}{(1+r)^t} \qquad (4.26)$$

The equation uses the absolute value of the negative space impacts (NNSV) and adds the negative external to the total for theoretical clarity.[5]

---

[5] See Papazian (2022) for additional examples and a more extensive discussion.

When an investment has no negative impacts, and its planetary, human, and economic impacts are all positive, investors may also consider adding its positive space impact as an optimisation target.

4.3.7.2   Space Impact Adjusted CAPM

When discussing the core equations of value and return in finance in Chapter 2, we observed that they are abstracted from space and do not consider the impact of investments or cash flows as an integral element of the return on investment. While the previous discussion looked at the NPV model, here I discuss the CAPM model in the context of impact adjusted returns.

To recap, the CAPM model defines the expected return on a security i as a function of its riskiness, using Beta, a volatility measure, as its proxy for risk.

$$R_i = R_f + \beta_i \times (R_m - R_f) \tag{4.27}$$

R_i = Return on security i
R_f = Risk Free Rate
β_i = Risk Proxy
R_m = Return on market

where Beta of the security is equal to:

$$\text{Beta}_i = \beta_i = \frac{\text{Covariance}_{R_i, R_m}}{\text{Variance}_{R_m}}$$

If the CAPM was to be considered within the space value framework, given the space value of money principle, the investment or security in question must be, at the very least, space neutral. In other words, its Net Space Value must be at least equal or greater than zero, and the same goes for its space growth rate.

Given that negative space impacts must be excluded, all those securities that have a negative impact cannot and should not be considered viable investments. Therefore, when we adopt the space value of money principle, the CAPM and many other APT and multi factor models become conditional models, where the condition is that the Net Space Value of the investment or security is either neutral or positive:

$$\text{NSV}_{T,S,i} \geq 0$$

$$\text{SPR}_i \geq 0$$

This implies that risk-adjusted returns can only be considered a meaningful pursuit for neutral/positive impact securities. For those investments that meet this condition and criteria, if the same risk/return logic were to be used to quantify the relationship between *impact and return*, i.e., investors can expect a higher return for additional positive impact, then we must define the space impact benchmark to formally express the relationship. Papazian (2022) offers a detailed discussion and proposes the below relationship:

$$R_i = R_{\min} + \text{MSP}_i \tag{4.28}$$

R_i = Return on Investment or Security i
$R_{\min}$ = R_{minimum} = Return on Investment with minimum ≥ 0 acceptable space impact
MSP_i = Market Space Premium on Security i

$$\text{MSP}_i = \theta_i \cdot (R_{\text{HIP}} - R_{\min})$$

R_{HIP} = Expected Return on High Space Impact Portfolio
θ_i = Theta = Space Impact Sensitivity of Security i

$$\theta_i = \frac{\text{Covariance}_{R_i, \text{NSV}_i}}{\text{Variance}_{\text{NSV}_i}}$$

NSV_i = Net Space Value of Security i

When considering space impact as a factor, not just as a condition for universe selection, then we can write:

$$R_i = R_f + \beta_i (R_m - R_f) + \theta_i (R_{\text{HIP}} - R_{\min}) \tag{4.29}$$

The actual relationship between positive space impact and returns is yet to be discovered. This can only become possible once the framework is adopted, the principle is applied, the specific space impact measures are calculated, and there is enough historical data to test and reveal the market sensitivity of returns. How much can investors require or expect to earn for each additional dollar of positive space impact will have to be the subject of future empirical research.

4.4   Conclusion

The value framework of finance theory and industry has been built to serve the mortal risk-averse return-maximising investor as only stakeholder, constructed around two key principles of value, time value of money and risk and return. This has led to models where the value of cash flows is measured and defined by risk and time parameters alone. As discussed in Chapter 2, our mathematics of value and return omits the analytical dimension of space, our physical context stretching from subatomic space to interstellar space and every layer in between and beyond. The space value of cash flows has not been part of the discussion, and thus, the responsibility of space impact has never been a defining element of the value of cash flows.

In Chapter 3, we observed that while growing at an impressive rate, sustainable finance and/or ESG integration has not yet penetrated the analytical content of core finance theory and practice. Indeed, despite numerous encouraging developments, sustainable finance frameworks, standards, scores, and tools do not go far enough to transform our equations of value and return.

This chapter presented the space value framework first introduced in Papazian (2022). Starting with the introduction of the analytical dimension of space, our physical context stretching from subatomic to interstellar space and every layer in between and beyond, the framework defines our relationship with space through the space value of money principle, which opens the door for the formal consideration of two new stakeholders in our models and equations of value and return, i.e., planet and humanity.

The space value of money principle complements our existing principles, i.e., time value of money and risk and return, and establishes the bottom threshold of investment acceptability by requiring that a dollar invested in space has at the very least a dollar's worth of positive impact on space. This ushers in an entirely new analytical framework where the detail mapping and quantification of the space impact of cash flows, and their integration into our equations, are central to financial analysis. The value of cash flows must now be based on more than their risk and time value. The quantification of impact and its integration into value begins by the expanded cash flow timeline, the double timeline, which also accounts for the space impact of cash flows. In other words, as investors discount their future expected cash flows to the present, they must now compound the space impact it would take to achieve or expect them into the future.

The chapter introduced the key concepts and equations of the space value framework: planetary, human, and economic impact, their corresponding intensity measures, their component elements, i.e., pollution, biodiversity, human capital, R and D, new asset, and new money impacts, and the aggregate measures Net Space Value, Gross Space Value, and the space growth rate. Moreover, two examples, the Impact Adjusted Net Present Value and the Impact Adjusted CAPM, described a plausible approach through which the space impact of cash flows can be integrated into our current models and analysis.

References

Bhagat, S., B Bolton, and R. Romano. (2008). The Promise and Peril of Corporate Governance Indices. *Colombia Law Review* 108 (8): 1903–1882. https://www.jstor.org/stable/40041812. Accessed 15 February 2022.

Bluedorn, A. 1982. Theories of Turnover: Causes, Effects, and Meanings. In *Research in the Sociology of Organisations*, ed. S.B. Bacharach, 75–128. Greenwich: JAI Press.

Brickley, J., and L. Drunen. 1990. Internal Corporate Restructuring: An Empirical Analysis. *Journal of Accounting and Economics* 12: 251–280. https://doi.org/10.1016/0165-4101(90)90050-E.Accessed02February2021.

Britannica. 2023. Pollution—Environment. Encyclopaedia Britannica. https://www.britannica.com/science/pollution-environment. Accessed June 2023.

Caldecott, B., N. Howarth, and P. McSharry. 2013. Stranded Assets in Agriculture: Protecting Value from Environment-Related Risks. Smith School of Enterprise and the Environment. https://www.smithschool.ox.ac.uk/publications/reports/stranded-assets-agriculture-report-final.pdf. Accessed 2 February 2021.

CGPM. 2022. Résolutions de la Conférence Générale des poids et mesures (27e réunion). Comité international des poids et mesures. https://www.bipm. org/documents/20126/64811223/Resolutions-2022.pdf/281f3160-fc56-3e63-dbf7-77b76500990f . Accessed 25 July 2023.

CPI. 2014. Moving to a Low-Carbon Economy: The Financial Impact of the Low Carbon Transition. Climate Policy Initiative. https://climatepolicyin itiative.org/wp-content/uploads/2014/10/Moving-to-a-Low-Carbon-Eco nomy-The-Financial-Impact-of-the-Low-Carbon-Transition.pdf. Accessed 2 February 2022.

Daines, R.M., I.D. Gow, and D.F. Larker. 2010. Rating the Ratings: How Good Are Commercial Governance Ratings? *Journal of Financial Economics* 98 (3): 439–461. https://doi.org/10.1016/j.jfineco.2010.06.005.Accessed0 2February2021.

Dasgupta, P. 2021. The Economics of Biodiversity: The Dasgupta Review. HM Treasury, London. https://assets.publishing.service.gov.uk/government/uploads/system/uploads/attachment_data/file/962785/The_Economics_of_Biodiversity_The_Dasgupta_Review_Full_Report.pdf. Accessed 2 March 2021.

Díaz, S., et al. 2019. Pervasive Human-Driven Decline of Life on Earth Points to the Need for Transformative Change. *Science* 366 (6471): eaax3100. https://doi.org/10.1126/science.aax3100. Accessed 2 February 2021.

Dietzel, A., and J. Maes. 2015. Costs of Restoration Measures in the EU Based on an Assessment of LIFE Projects. European Commission. https://public ations.jrc.ec.europa.eu/repository/bitstream/JRC97635/lb-na-27494-en-n. pdf. Accessed 12 January 2020.

Dissanaike, G., A. V. Papazian. 2005. Management Turnover in Stock Market Winners and Losers: A Clinical Investigation. European Corporate Governance Institute. ECGI. Finance Working Paper N° 61/2004. https://papers.ssrn.com/sol3/papers.cfm?abstract_id=628382. Accessed 2 February 2021.

Eames, C., Eames, R. 1977. Powers of Ten. Eames Office. http://www.powers of10.com/. Accessed 14 June 2023.

EIA. 1995. Electricity Generation and Environmental Externalities: Case Studies. *Energy Information Administration.* https://www.nrc.gov/docs/ML1402/ML14029A023.pdf. Accessed 02 February 2022.

Ellerman, A.D., J.F. Convery, and C. De Perthius. 2010. *Pricing Carbon: The European Union Emissions Trading Scheme.* Cambridge: Cambridge University Press.

Environment Agency. 2015. Cost Estimation for Habitat Creation—Summary of Evidence. Environment Agency. https://assets.publishing.service.gov.uk/media/6034ef5ee90e0766033f2ea7/Cost_estimation_for_habitat_creation. pdf. Accessed 2 June 2020.

ESA. 2017. Biodiversity. Ecological Society of America. https://www.esa.org/wp-content/uploads/2012/12/biodiversity.pdf. Accessed 2 February 2022.

Fama, E.F. 1980. Agency Problems and the Theory of the Firm. *Journal of Political Economy*, 660–672. https://www.jstor.org/stable/1837292. Accessed 2 February 2021.

Fama, E.R., and M.C. Jensen. 1983. Separation of Ownership and Control. *Journal of Law and Economics*, 301–325. https://www.jstor.org/stable/725104. Accessed 2 February 2021.

Gilson, S. 1989. Management turnover and financial distress. *Journal of Financial Economics* 25: 241–262. https://doi.org/10.1016/0304-405 X(89)90083-4.Accessed02February2021.

Global Canopy, Vivid Economics. 2020. The Case for a Task Force on Nature-Related Financial Disclosures. https://globalcanopy.org/wp-content/upl oads/2020/11/Task-Force-on-Nature-related-Financial-Disclosures-Full-Rep ort.pdf. Accessed 12 December 2021.

Halpern, B.S., et al. 2015. Spatial and Temporal Changes in Cumulative Human Impacts on the World's Ocean. *Nature Communications* 6 (1): 1–7. https://doi.org/10.1038/ncomms8615.

Hawking, S. 2008. Why We Should Go into Space. National Space Society. https://space.nss.org/stephen-hawking-why-we-should-go-into-space-video/. Accessed 2 February 2022.

HM Government. 2021. Greenhouse Gas Reporting: Conversion Factors 2021. UK Government. https://www.gov.uk/government/publications/gre enhouse-gas-reporting-conversion-factors-2021. Accessed 2 February 2022.

HM Government. 2019. Government Greenhouse Gas Conversion Factors for Company Reporting: Methodology Paper for Emission Factors Final Report. UK Government. Department for Business, Energy, and Industrial Policy. https://assets.publishing.service.gov.uk/government/uploads/system/upl oads/attachment_data/file/904215/2019-ghg-conversion-factors-method ology-v01-02.pdf. Accessed 2 February 2020.

HM Government. 1994. Biodiversity: The UK Action Plan. Her Majesty's Stationery Office (HMSO), Office of Public Sector Information, UK. https://data.jncc.gov.uk/data/cb0ef1c9-2325-4d17-9f87-a5c84fe400bd/ UKBAP-BiodiversityActionPlan-1994.pdf. Accessed 2 February 2021.

Hossenfelder, S. 2012. Can We Measure Structures to a Precision Better Than the Planck Length? *Classical and Quantum Gravity* 29 (11). https://doi.org/10.1088/0264-9381/29/11/115011. Accessed 2 August 2023.

Hooke, R.L., M.J.F. Duque, and J.D. Pedraza. 2012. Land Transformation by Humans: A Review. *GSA Today* 22: 4–10. https://www.geosociety.org/gsa today/archive/22/12/pdf/i1052-5173-22-12-4.pdf. Accessed 2 February 2021.

HSBC-BCG. 2021. Delivering Net Zero Supply Chains: The Multi-Trillion Dollar Key to Beat Climate Change. Boston Consulting Group and HSBC. https://www.hsbc.com/-/files/hsbc/news-and-insight/2021/pdf/211026-delivering-net-zero-supply-chains.pdf?download=1. Accessed 2 February 2022.

Ihrig, J., Weinbach, G.C., Wolla, S.A. 2021. Teaching the Linkage Between Banks and the Fed: R.I.P. Money Multiplier. Econ Primer. Federal Reserve Bank of St. Louis. https://research.stlouisfed.org/publications/page1-econ/2021/09/17/teaching-the-linkage-between-banks-and-the-fed-r-i-p-money-multiplier. Accessed 02 January 2022.

IPBES. 2019. The Global Assessment Report on Biodiversity and Ecosystem Services. Intergovernmental Science-Policy Platform on Biodiversity and Ecosystem Services. https://ipbes.net/system/files/2021-06/2020%20IPBES%20GLOBAL%20REPORT%28FIRST%20PART%29_V3_SINGLE.pdf Accessed 2 February 2021.

IPCC. 2022. Climate Change 2022: Impacts, Adaptation and Vulnerability. Summary for Policymakers. Intergovernmental Panel on Climate Change. https://report.ipcc.ch/ar6wg2/pdf/IPCC_AR6_WGII_SummaryForPolicymakers.pdf. Accessed 28 February 2022.

IPCC. 2021. Climate Change 2021: The Physical Science Basis. IPCC. https://www.ipcc.ch/report/ar6/wg1/downloads/report/IPCC_AR6_WGI_SPM_final.pdf. Accessed 2 February 2022.

IPCC. 2018. Summary for Policymakers. In: Global Warming of 1.5°C. An IPCC Special Report on the Impacts of Global Warming of 1.5°C Above Pre-industrial Levels and Related Global Greenhouse Gas Emission Pathways, in the Context of Strengthening the Global Response to the Threat of Climate Change, Sustainable Development, and Efforts to Eradicate Poverty. IPCC. https://www.ipcc.ch/site/assets/uploads/sites/2/2019/05/SR15_SPM_version_report_LR.pdf. Accessed 2 February 2022.

IPCC. 2013. Climate Change 2013: The Physical Science Basis. Summary for Policymakers. Intergovernmental Panel on Climate Change. https://www.ipcc.ch/site/assets/uploads/2018/03/WG1AR5_SummaryVolume_FINAL.pdf. Accessed 2 February 2021.

Jensen, M.C., and W.H. Meckling. 1976. Theory of the Firm: Managerial Behaviour, Agency Costs, and Ownership Structure. *Journal of Financial Economics*, 305–360. https://doi.org/10.1016/0304-405X(76)90026-X.Accessed02February2021

Krausmann, F., et al. 2017. Global Socioeconomic Material Stocks Rise 23-Fold Over the 20th Century and Require Half of Annual Resource Use. *Proceedings of the National Academy of Sciences of the United States of America* 114 (8): 1880–1885. https://doi.org/10.1073/pnas.1613773114.

Maxwell, S.L., R.A. Fuller, T.M. Brooks, and J.E. Watson. 2016. Biodiversity: The Ravages of Guns, Nets and Bulldozers. *Nature News* 536 (7615): 143–145. https://doi.org/10.1038/536143a.

McLeay, M., Radia, A., Thomas, R. 2014. Money creation in the modern economy. *Bank of England Quarterly Bulletin*. https://www.bankofengland. co.uk/-/media/boe/files/quarterly-bulletin/2014/money-creation-in-the-modern-economy. Accessed 06 June 2022.

NPL. 2023. NPL Leads Expansion to the SI Prefix Range for the Global Metrology Community. National Physical Laboratory. https://www.npl.co. uk/si-prefix. Accessed 27 July 2023.

OECD. 2021. *Effective Carbon Rates 2021: Pricing Carbon Emissions Through Taxes and Emissions Trading*. Paris: OECD Publishing. https://doi.org/10. 1787/0e8e24f5-en.Accessed12February2022

OECD. 2018. *Effective Carbon Rates 2018: Pricing Carbon Emissions Through Taxes and Emissions Trading*. Paris: OECD Publishing.

OECD. 2015. *Frascati Manual 2015: Guidelines for Collecting and Reporting Data on Research and Experimental Development, The Measurement of Scientific, Technological and Innovation Activities*. Paris: OECD Publishing. https://doi.org/10.1787/9789264239012-en.Accessedon02February2022

Padmanabhan, T. 1985. Physical Significance of Planck Length. *Annals of Physics* 165: 38–58. https://doi.org/10.1016/S0003-4916(85)80004-X.Accessed3 0July2023.

Papazian, Armen. 2022. *The Space Value of Money: Rethinking Finance Beyond Risk and Time*. New York: Palgrave Macmillan. https://doi.org/10.1057/ 978-1-137-59489-1.

Papazian, A.V. 2004. An Endoscopy on Stock Market Winners and Losers. Unpublished PhD Dissertation. Cambridge University Judge Business School Library. Cambridge, UK

Parker, D.M. 1995. Habitat Creation—A Critical Guide. *English Nature Science* 21. http://publications.naturalengland.org.uk/publication/2294780 Accessed 2 February 2021

Pearce, D., and D. Moran. 1994. *The Economic Value of Biodiversity. International Union for the Conservation of Nature—The World Conservation Union*. Taylor and Francis, UK.

Seiberg, N. 2006. *Emergent Spacetime*. School of Natural Sciences, Institute for Advanced Study. Princeton. https://arxiv.org/pdf/hep-th/0601234.pdf. Accessed 30 March 2023.

SETAC. 1991. A Technical Framework for Life Cycle Assessment. *Society of Environmental Toxicology and Chemistry*. https://cdn.ymaws.com/www.setac. org/resource/resmgr/books/lca_archive/technical_framework.pdf. Accessed 02 February 2022.

SETAC. 1993. Guidelines for Life-Cycle Assessment: A "Code of Practice". *Society of Environmental Toxicology and Chemistry.* https://cdn.ymaws.com/www.setac.org/resource/resmgr/books/lca_archive/guidelines_for_life_c ycle.pdf. Accessed 02 February 2022.

SFPUO. 2018. Climate Risk Analysis from Space: Remote Sensing, Machine Learning, and the Future of Measuring Climate-Related Risk. Sustainable Finance Programme. https://www.smithschool.ox.ac.uk/research/sustai nable-finance/publications/Remote-sensing-data-and-machine-learning-in-cli mate-risk-analysis.pdf. Accessed 2 February 2021.

Shepherd, P., B.S. Gillespie, and D. Harley. 1999. Preparation and Presentation of Habitat Replacement Costs Estimates. *English Nature* 345. http://public ations.naturalengland.org.uk/publication/63034. Accessed 2 February 2021.

Shleifer, A., and R. Vishny. 1989. Managerial Entrenchment: The Case of Firm-Specific Assets. *Journal of Financial Economics* 25: 123–139.

Shleifer, A., and R. Vishny. 2012. A Survey of Corporate Governance. *The Journal of Finance* 52 (2): 737–783.

Smolin, L. 2006. *The Trouble with Physics.* England: Penguin Books.

Stein, J.C. 1989. Efficient Capital Markets, Inefficient Firms: A Model of Myopic Corporate Behavior. *Quarterly Journal of Economics* 104: 655–669.

Tittensor, D.P., et al. 2014. A Mid-term Analysis of Progress Toward International Biodiversity Targets. *Science* 346 (6206): 241–244. https://doi.org/10.1126/science.1257484.

Trucost. 2013. *Natural Capital at Risk: The Top 100 Externalities of Business.* London: Trucost.

UNFCCC. 2021. *COP 26 and the Glasgow Financial Alliance for Net Zero (GFANZ).* https://racetozero.unfccc.int/wp-content/uploads/2021/04/GFANZ.pdf. Accessed 2 February 2022.

UNFCCC. 2015. Paris Agreement. United Nations Framework Convention on Climate Change. https://unfccc.int/sites/default/files/english_paris_agre ement.pdf. Accessed 2 December 2020.

Venter, O., et al. 2016. Sixteen Years of Change in the Global Terrestrial Human Footprint and Implications for Biodiversity Conservation. *Nature Communications* 7 (1): 1–11. https://doi.org/10.1038/ncomms12558.

WEF. 2021. Global Risks Report 2021. https://www.weforum.org/reports/the-global-risks-report-2021.

White, B.T., L.R. Viana, G. Campbell, C. Elverum, and L.A. Bennun. 2021. Using Technology to Improve the Management of Development Impacts on Biodiversity. *Business Strategy and the Environment* 30: 3502–3516. https://doi.org/10.1002/bse.2816.

World Bank. 2020. The Human Capital Index: 2020 Update. The World Bank, Washington. https://openknowledge.worldbank.org/handle/10986/34432. Accessed 2 February 2021.

WRAP. 2018. Gate Fees 2017/18 Final Report: Comparing the Costs of Alternative Waste Treatment Options. WRAP. http://www.wrap.org.uk/sites/files/wrap/WRAP%20Gate%20Fees%202018_exec+extended%20summary%20report_FINAL.pdf. Accessed 26 December 2018.

WWF. 2018b. Living Planet Report—2018: Aiming Higher, ed. M. Grooten and R.E.A. Almond. WWF. https://www.worldwildlife.org/pages/living-planet-report-2018.

CHAPTER 5

Implications for Money Mechanics

Abstract This chapter discusses the implications of adopting the space value framework for money mechanics. While hardwiring sustainability into financial mathematics based on the space value of money principle and ensuing equations will require a comprehensive reassessment of instruments and transactions used by money creators, i.e., central and commercial banks, this chapter focuses on the broader systemic considerations given our debt-based monetary architecture. It discusses three major limitations of debt-based money, calendar time, monetary gravity, and monetary hunger, and offers a new money creation logic based on space value creation. It proposes Value Easing through Public Capitalisation Notes as a transactional approach that can also facilitate the funding of the many trillions of dollars needed for the transition to a Net Zero sustainable economy.

Keywords Sustainability · Financial Mathematics · Money · Value · Risk · Time · Space · Impact

JEL Classification E00 · E58 · G00 · G30 · Q51

A. V. Papazian, *Hardwiring Sustainability into Financial Mathematics*, https://doi.org/10.1007/978-3-031-45689-3_5

If investors must respect the space value of money principle, and earn their returns with a positive space impact, then money creators, whether they are commercial or central banks, must also follow suit. This means that every instance of contractual financial engineering that results in new money creation must apply the same financial mathematics, where the value/return of the instruments must be considered not just in terms of time and risk, but also space, and space impact.

Indeed, one of the key contributions of this book is to rectify another important omission in sustainable finance. As explored and discussed in Chapter 3, the many frameworks, standards, scores, and tools that aim to operationalise sustainability in finance seem to treat money and the logic of its creation as exogenous to the sustainability challenge/opportunity—as if, money, and the principles and equations that govern its creation, allocation, and deployment are not material to the sustainability effort that aims to reinvent human productivity.

The hardwiring of sustainability into financial mathematics as discussed in Chapter 4 will naturally require a comprehensive reassessment of instruments and transactions used by money creators. While this in itself will have a variety of technical implications for the asset portfolios of commercial and central banks, in this chapter I focus on broader systemic considerations given our debt-based monetary architecture.

5.1   DEBT-BASED MONEY

In a paper published in the Bank of England Quarterly Bulletin, our current debt-based architecture is best described by McLeay et al. (2014a):

> There are three main types of money: currency, bank deposits and central bank reserves. Each represents an IOU from one sector of the economy to another. Most money in the modern economy is in the form of bank deposits, which are created by commercial banks themselves. (McLeay et al. 2014a, 4)

Table 5.1 provides examples of instruments, portfolios, and transactional engagements that lead to or are used for debt-based money creation within our current architecture. Figure 5.1 elaborates the debt/loan-based money mechanics structured around IOU (I Owe You) transactions/instruments between central banks, commercial banks, and

Table 5.1 Commercial and Central Bank sample debt instruments, portfolios, and transactional engagements

	Commercial banks	*Central banks*
Instruments	Consumer Credit Business Credit Residential Mortgages Commercial Mortgages	Discount Loans (FED) TLTRO Loans (ECB) Subsidiary Loans (BOE)
Portfolios	Loan Portfolios Mortgage Portfolios	Government Bond Portfolios Corporate Bond Portfolios MBS and CDO Portfolios Commercial Paper (FED)
Transactional Engagements	Loan Approvals Mortgage Approvals	Currency Issuance Reserve Issuance - Quantitative Easing - Credit Easing

Source Author

consumers. Chart 5.1 depicts the Asset side of the Federal Reserve balance sheet denoting key QE/CE periods of money creation.

5.2 Challenges with Debt-Based Money

When we consider our debt-based money mechanics within the context of a transformed analytical value framework and a space-adjusted financial mathematics of value and return, a number of key challenges become evident. Specifically, three systemic dependencies and limitations reveal the necessity to rethink our monetary architecture for a true change in trajectory that can actually deliver the sustainability we seek.

    a. Calendar time
    b. Monetary gravity
    c. Monetary hunger

I briefly discuss and summarise these challenges next.

5.2.1   Calendar Time

Creating money based on debt instruments involves calendar time obligations. While this is taken for granted within our monetary and financial

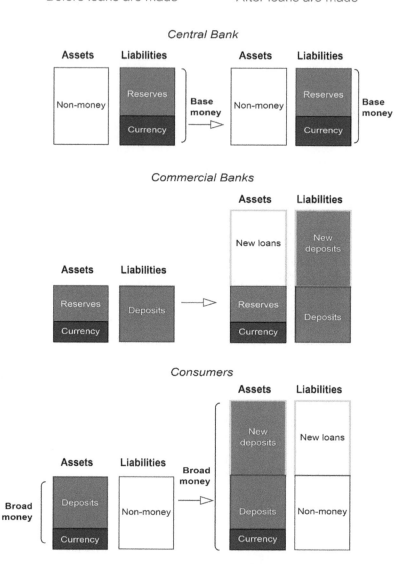

Fig. 5.1 Bank of England money creation process through loans (*Source* Adapted from McLeay et al. [2014b])

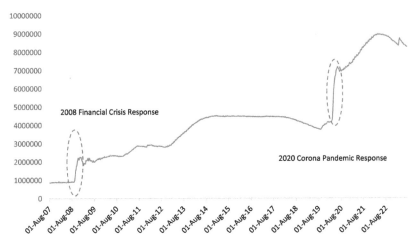

Chart 5.1 Federal Reserve balance sheet, assets in $ millions (*Source* FED [2022])

architecture, in truth, linking money creation to an artificial concept like calendar time poses serious constraints. Due to this calendar time-based conceptualisation of money creating instruments, governments, government agencies, municipalities, small businesses, households, individuals, corporations, and even banks are all chained to calendar time payments.

> Debt, which involves time obligations in terms of scheduled interest and principal repayments, chains everybody involved to calendar time payments. Indeed, whatever the actual shape of the repayment schedule involved, our debt-based money creation methodology chains our entire productive and creative potential to calendar time. (Papazian 2022, 214)

It is very important to note here that I do not use the term time, but calendar time. While the nature of time is a debatable subject (Rovelli, 2018; Smolin 2006; Greene 2004) and economists have discussed the relevance of psychological and real time to the performance of investments (Blanqué 2021), the time used and applied by banks and central banks in debt transactions is simply calendar time.

Calendar time is central to the functioning of the world economy. Calendar time gives structure and direction to our days and our productive activities, and it is a human invention that helps us structure and

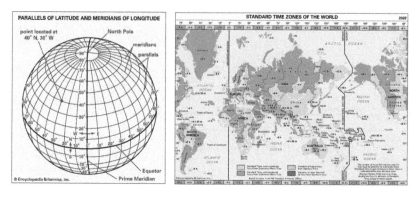

Fig. 5.2 Prime Meridian and standard time zones (*Source* Britannica [2023a, b])

navigate our productive life on the planet. The conceptual mapping of space and time on Earth makes use of the Prime Meridian, epitomised by the Greenwich laser beam, at 0° longitude (See Fig. 5.2).

> The Prime Meridian is the line and the point at which the world's longitude is set at 0°. It does not exist in any strict material sense, yet through maps and clocks, the prime meridian governs the life of every human on Earth. (Withers 2017, 5)

While this imaginary point/line on Earth, in space, is an important element and structural pillar of the entire world economy, using an artificial point or line and the resulting fixed-paced calendar time as a foundation of money mechanics limits our ability to invest in space time-lessly beyond the limitations of the structural construct. This is one of the key issues with calendar time and debt-based money.

5.2.2   *Monetary Gravity*

Debt-based money creation, at the central bank level and at the commercial bank level, is achieved through loans and debt instruments that require the repayment of principal and interest to the original source within a specific calendar time window. In other words, debt-based money involves a backward loop to the creator of money within specified calendar time windows and intervals. While this is taken for granted in our current systems, it is also the source of a unique type of artificially created force that I call monetary gravity (Papazian 2022, 216).

Debt-based money is a human invention, and it acts as a powerful constraint on how far in space an investment can reach before it must return to pay back principal and interest to some bank. Indeed, this is true even if payments are done electronically without a physical return to the bank because the structure of the instrument imposes the necessity to earn the income and make the payment within the time frame required. Even if the debt can be rolled over, or postponed, and refinanced, the foundation of debt-based money imposes a limit on how far in space a process can go before having to return to some bank.

The below conceptual equation can be used to calculate the limit on distance travelled when a monthly interest payment must be paid on a money creating instrument. Table 5.2 provides the limits on distances given the speed of Usain Bolt, the SSC Tuatara, the Parker Probe (NASA 2018), and light.[1]

$$\text{Maximum Distance}_{\text{Light}} = \text{Speed of Light} \frac{m}{s} \times \text{Time Interval in } s$$

Table 5.2 Distance travelled in a month in m

	Usain Bolt	SSC Tuatara	Parker Solar Probe	Light
Distance (m)	27,060,480	366,158,880	177,811,397,406.72	777,062,051,136,000

Source Author, from Papazian (2022)[2]

[1] Please note that the calculations assume uniform terrestrial conditions for simplicity and the purpose of the argument.

[2] Calculation details: the distance light can travel in one month, also known as a light-month, is the distance that light travels in an absolute vacuum in one full month. The speed of light is equal to 299 792 458 m/s, assuming 30 days in a month, and 86,400 seconds in each, in one month light travels 777,062,051,136,000 metres, which is equivalent to approximately 777 Tm (1 Terametre = 1,000,000,000,000 metres). Usain Bolt, the Jamaican sprinter, set the world record in 2009 in the 100-metre sprint at 9.58 seconds, giving him a speed of 10.44 metres per second, which means the furthest Bolt can run in one month is 27,060,480 metres. The fastest production, the SSC Tuatara, reported to have a speed of 316 miles per hour, or 141.265 metres per second, the furthest SSC Tuatara can travel in a month is 366,158,880 metres. The Parker Solar

These are hypothetical examples in order to argue that we and our fastest tools and inventions, and light itself, will experience a limit on distance travelled given an interval of calendar time.

> Debt-based money acts as a leash on our species, chaining us to a self-created calendar, to a self-created system that ultimately chains us to the surface of the planet. Given that calendar time is a human concept, artificially created to manage human activities, linking money creation to an artificially limited concept such as a month or a year, artificially limits the distance we can travel before we need to return to the bank. (Papazian 2022, 218)

5.2.3   Monetary Hunger

The third key space impact challenge posed by debt-based money is what I call monetary hunger. In any debt-based economy, and at any point in time, irrespective of past or current capital accumulation, a large segment of society, including households, municipalities, governments, corporations, and banks, is chasing available cash and deposits to pay calendar time-linked debt obligations. Debt-based money creates this chase, this monetary hunger, in any debt-based economy irrespective of the actual levels of debt to GDP ratios. This is so given that money is continuously created via debt.

As a numerical example of monetary hunger, in the United States, between 2000 and 2022, the total outstanding public and private debt across all sectors increased from 28.7 trillion US dollars to 93.5 trillion US dollars (See Chart 5.2).

While a significant driver of growth, monetary hunger can also explain personal, business and investment practices. After all, the threat of default and loss of ratings and assets are existential threats for businesses and households and avoiding them is naturally a priority. In other words, given voluntary sustainability frameworks and standards, and the threat of actual default, businesses will always choose to serve their debts before the environment.

Our debt-based monetary architecture is given primacy over our ecosystem—an absurdity with certain evolutionary implications

---

Probe (NASA 2018) achieved speeds of 153,454 miles per hour or 68,600.076 metres per second, the furthest it can travel within a month is 177,811,397,406.72 metres.

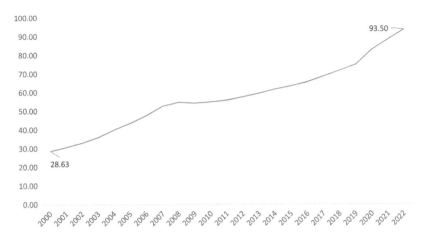

Chart 5.2 Total outstanding public and private debt USA, 2000–2022 in trillion US dollars (*Source* Statista [2022])

and potentially catastrophic consequences that is often legitimised by some economic belief or theory.

The omission of money mechanics from the discussion in sustainable finance is revealing—our frameworks and standards are focused on companies and corporations and are unconcerned with the very logic and process of money creation. This is so in spite and despite the sustainable banking initiatives which, once again, are aimed at their portfolios and emissions rather than the very logic of money creation. Our debt-based monetary architecture and the interests it represents have legally enforceable primacy over our voluntary frameworks and standards aimed at achieving planetary sustainability and securing the health of our only ecosystem.

Before presenting an alternative money creation logic that integrates space value and addresses the above discussed challenges, it is necessary to briefly discuss the relatively recent phenomenon of cryptocurrencies, or cryptoassets.

5.3   CRYPTOCURRENCIES

It is important to open this parenthesis on cryptocurrencies[3] given the popular misconception about their relevance and importance as an alternative form of money (to debt-based fiat money). Indeed, this is particularly relevant from a sustainable finance perspective.

As of the 7th of August 2023, there were 9752 cryptocurrencies listed on coinmarketcap.com (CoinMarketCap 2023). Bitcoin is the first on the list and is used as an example in this discussion without elaborating on the technical or other differences between and among all the listed cryptocurrencies. There are many features of Bitcoin that require attention. The first is their labelling as currency. The Bank of England refers to them as cryptoassets:

> Put it this way, you wouldn't use cryptocurrency to pay for your food shop. In the UK, no major high street shop accepts cryptocurrency as payment. It's generally slower and more expensive to pay with cryptocurrency than a recognised currency like sterling. Development is underway to make cryptocurrency easier to use, but for now it isn't very 'money-like'. This is why central banks now refer to them as 'cryptoassets' instead of 'cryptocurrencies'. Today cryptocurrencies are generally held as investments by people who expect their value to rise. (Bank of England 2020)

This raises an equally serious concern, they do not have any intrinsic value, and they do not involve any return accruing to holders. Prasad (2021) refers to this by referencing the 'greater fool' theory. He writes: "The valuations of meme currencies seem to be based entirely on the 'greater fool' theory—all you need to do to profit from your investment is to find an even greater fool willing to pay a higher price than you paid for the digital coins." An alternative interpretation is that the attractiveness of Bitcoin and other cryptocurrencies is in their lack of transparency, as 'dark money' (Economist 2022).

Another critical aspect is the energy consumption attached to the mining process that generates the coins. The Cambridge Bitcoin Energy Consumption Index (CBECI) estimates that the yearly average annualised electricity consumption of the Bitcoin Network is around 118.3 TWh per

---

[3] Cryptocurrencies should not be confused with Central Bank Digital Currencies (CBDC) which are now being considered by the Bank of England and other central banks (Bank of England 2023a).

year, which is higher than the yearly consumption of the Netherlands at 113.3 TWh per year (CCAF 2023).[4] Moreover, a recent study reveals the electronic waste generated due to the consumption of electronic hardware used in mining the coins. The authors estimate that Bitcoin's e-waste is 30.7 metric kilotons per annum as of May 2021 (De Vries and Stoll 2021).

While the above raised issues are all relevant and important to sustainability and sustainable finance, the most critical, however, is the very logic of their creation. While central bank and commercial bank money are created via debt instruments/transactions, crypto coins are created after a mining process through powerful computers performing specific mining operations which consist in solving complex mathematical puzzles (Bitcoin 2021a, b, c).

> Anybody can become a Bitcoin miner by running software with specialized hardware. Mining software listens for transactions broadcast through the peer-to-peer network and performs appropriate tasks to process and confirm these transactions. Bitcoin miners perform this work because they can earn **transaction fees paid by users for faster transaction processing, and newly created bitcoins issued into existence according to a fixed formula.**

> For new transactions to be confirmed, they need to be included in a block along with a mathematical proof of work. **Such proofs are very hard to generate because there is no way to create them other than by trying billions of calculations per second.** This requires miners to perform these calculations before their blocks are accepted by the network and before they are rewarded. As more people start to mine, the difficulty of finding valid blocks is automatically increased by the network to ensure that the average time to find a block remains equal to 10 minutes. As a result, mining is a very competitive business where no individual miner can control what is included in the block chain (Bitcoin 2021a).[5]

The underlying process through/for which Bitcoins are created/awarded involves and is dependent on *'trying billions of calculations per second.'*

---

[4] For a relative understanding of the numbers, note that the two highest-consuming countries, China and the United States, use respectively 7805.66 TWh and 3979.28 TWh per year (CCAF 2023).

[5] Emphasis added.

In other words, to put it in the most benign way possible, minors get rewarded with bitcoin for mathematical guesswork. Basically, from a debt logic, we have now moved to a more preposterous logic of money creation, mathematical guesswork.

The above discussion aimed at clarifying an important aspect of the debate, and to remove all doubts regarding the popular misconception that cryptocurrencies are an actual alternative to debt-based money. From a sustainability perspective, they in fact offer no real solutions to the challenges discussed in the previous sections, and in truth, they add a host of new ones that further undermine a sustainable trajectory.

5.4   Money Mechanics with Space Value Creation

Previous sections discussed the key challenges of debt-based money from a systemic perspective, and revealed why cryptocurrencies are not a sustainable alternative we can rely on. Indeed, from debt to mathematical guesswork with a heavy carbon and electronic footprint, we seem to be in dire need of a money mechanics that is built on and delivers positive space impact.

This section elaborates on an alternative money creation logic and instrument that are aligned with the space value of money principle and can provide the blueprint for a mechanism that can fund the transition to Net Zero, and help us address the many evolutionary challenges we have created for ourselves. Given the many trillions of dollars we will need to fund the transition (HSBC-BCG 2021; McKinsey-GI 2022), and the absence of a reliable framework to create and deploy these funds, this proposition is of critical relevance.

It is important to start by stating why it is and should be possible to improve and fine-tune our money creation methodology. Indeed, it is very much a plausible opportunity and necessity to introduce a new channel of money creation that is not debt-based, and addresses the challenges discussed in the previous sections. The below quote and Table 5.3 reveal why we should and must consider this avenue as a veritable opportunity for innovation and improvement in money mechanics.

[I]f the Bank of England can create and back banknotes by a deposit in the banking department of the Bank of England, if the Bank of England can create new money by loaning to its own wholly owned subsidiary, if the Federal Reserve can create new money by buying toxic Collateralised

Debt Obligations and Mortgage-Backed Securities or by buying commercial paper, there is no reason why they cannot back or create new money through an alternative equity-like instrument that shares risks, shares the ownership of the assets created through the instrument, has a tangible and inspiring positive space impact, and helps resolve our evolutionary challenges. (Papazian 2022, 223)

Table 5.3 depicts the balance sheet of the issue department of the Bank of England and reveals that much of the British Pounds in circulation are backed by an internal deposit at the banking department of the Bank of England. Moreover, as the discussion in Papazian (2022) illustrates, during the 2008 financial crisis and the 2020 Coronavirus pandemic, the Bank of England injected hundreds of billions of new money into the financial system through loans to its own wholly owned subsidiaries using the Asset Purchase Facility (APF) created by HM Treasury.

When the APF is used for monetary policy purposes, purchases of assets are financed by the creation of central bank reserves. The APF transactions are undertaken by a subsidiary company of the Bank of England—the Bank of England Asset Purchase Facility Fund Limited (BEAPFF). The transactions are funded by a loan from the Bank... (Bank of England 2021, 117)

The idea put forward here is quite straightforward and amounts to the introduction of a new financial instrument within our existing monetary architecture. The proposition is to introduce a new instrument of money creation, Public Capitalisation Notes (PCNs), which have a different logic and different locus of injection from previously used debt instruments through quantitative and/or credit easing (Bernanke 2009) by the Federal Reserve, Bank of England, and the European Central Bank.

5.4.1 Public Capitalisation Notes (PCNs)

Public Capitalisation Notes are conceived as instruments that can be used by any central bank to inject new money into the economy. A number of key features make PCNs very different from conventional debt instruments. These features are designed to address the challenges discussed with debt-based money and introduce a new logic of money creation and injection founded on space value creation. If the trigger of money creation in debt transactions is the agreement to repay, in PCNs, the trigger is the commitment to create necessary positive space value and share the

Table 5.3 BOE balance sheet, issue department, in (£mn)

	2023 £mn	2022 £mn
Assets		
Securities of, or guaranteed by, the British Government	1,536	1,698
*Other securities and assets including those acquired under reverse repurchase agreements	84,371	84,742
Total Assets	**85,907**	**86,440**
Liabilities		
Note Issued		
In Circulation	85,907	86,440
Total Liabilities	**85,907**	**86,440**
Other securities and assets including those acquired under reverse repurchase agreements		
Deposit with Banking Department	84,261	82,387
Reverse repurchase agreements	110	2,355
	84,371	**84,742**

Source Bank of England (2023b)

returns when they occur (See Fig. 5.3 for injection flow and Fig. 5.4 for an example Climate PCN).

Figure 5.3 expands the rationale behind the change of locus of the injection. PCNs do not condition the space impact of the created new money by further or next round lending by banks. The injected liquidity does not start in the form of bank reserves and is therefore deposit/spending before it becomes bank reserves. In the case of Quantitative Easing and Credit Easing, the starting point is bank reserves and credit conditions, and the subsequent availability and direction of credit are dependent on bank lending.

As discussed above, introducing a new logic of money creation through the introduction of a new instrument that can be used in parallel allows systemic flexibility and the opportunity to address many of the bottle-necks we face. Most importantly, it allows the funding of evolutionary challenges and economy-wide investment programmes that cannot be sustained through debt financing alone and are at odds with the risk and time focused priorities of banks.

It is important to note that the suggestion here is not to 'print' or create new money indiscriminately. The proposition is to change the logic of creation for the amounts already being created in order to balance the

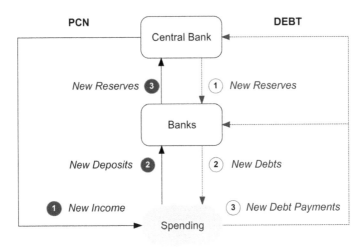

Fig. 5.3 PCN vs Debt money injection (*Source* Adapted from Papazian [2022])

system and address the challenges of debt-based money. I call this new approach of monetisation Value Easing (VE).

5.4.2   Value Easing

Value Easing: The transactional process undertaken by a central bank that consists in purchasing non-debt no-maturity equity-like high space impact value creating instruments from qualified government agencies and/or public private partnerships (PPP) with relevant Treasury sponsorship that increases the central bank's balance sheet and injects new liquidity outside the banking system/reserves.

Note that VE is far less inflationary than the recently used Quantitative Easing and Credit Easing, and it allows the injection to be directed where it is most needed, thus alleviating business cycle pressures, investment gaps, evolutionary challenges, and other strategic objectives. It addresses productivity and output gaps and enables the targeted injection of new liquidity that can alleviate supply shock-driven price inflation without relying on further bank lending—which, as previous experience has demonstrated post-2008 and 2020, can lead to asset price inflation rather than productive output growth.

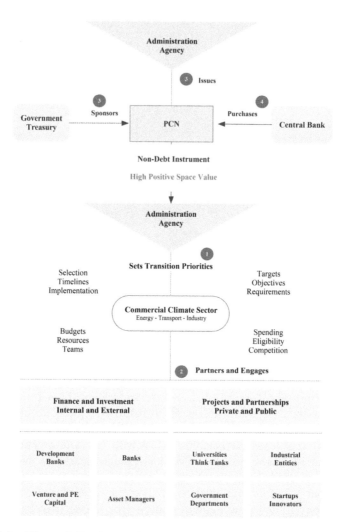

Fig. 5.4 Climate PCN (*Note* The key features of PCNs: Non-debt [addresses monetary hunger], no-maturity [addresses monetary gravity], equity-like [shares risks and assets and returns when due], high space value/impact [addresses challenges like transition], issued by qualified government agencies [as administrative HQ], in collaboration with private sector [Public Private Partnerships], with Treasury sponsorship [shares risk and assets and returns when due], increase the central bank's balance sheet [like QE and CE], inject new liquidity outside the banking system/reserves [unlike QE and CE]. *Source* Adapted from Papazian [2022])

Indeed, Value Easing is perfectly suited to fund those economy-wide transformations that cannot be appropriately addressed by our current risk and time-focused models and instruments.

5.4.3   From Debt Ceiling to Wealth Floor

Another brief parenthesis is appropriate here given the role of the US economy and the US dollar in the daily functioning of the global financial system. Despite ambitions to compete and/or replace the US dollar, as of today, the dollar remains the most commonly used global reserve currency (IMF 2023) at 54.66% of total (Chart 5.3).

Global markets were rattled by the intense debate around the US debt ceiling in the first half of 2023. The US Treasury reached its debt ceiling of $31.4 trillion in January 2023 (Edelberg and Sheiner 2023) and the ceiling was raised on the 31st of May 2023 after political brinkmanship across the aisle in the US Congress.

The debt limit or ceiling has been a central institutional bottleneck and the subject of intense debate in the United States (Austin 2015). The US Department of the Treasury describes the Debt Limit/Ceiling as follows:

The debt limit is the total amount of money that the United States government is authorized to borrow to meet its existing legal obligations,

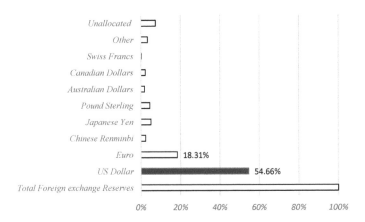

Chart 5.3   Global currency reserves percentages (IMF 2023)

including Social Security and Medicare benefits, military salaries, interest on the national debt, tax refunds, and other payments. (US Treasury 2023)

A few weeks before the new bill was passed by Congress, on the 3rd of May 2023, in a blog that reads as a cautionary tale, the White House writes:

> New analyses by both the Congressional Budget Office and the U.S. Department of the Treasury suggest the United States is rapidly approaching the date at which the government can no longer pay its bills, also known as the "X-date." History is clear that even getting close to a breach of the U.S. debt ceiling could cause significant disruptions to financial markets that would damage the economic conditions faced by households and businesses. Real time data, shown below, indicate that markets are already pricing in political brinkmanship related to Federal government default through higher risk premia. (White House 2023)

The x-date is the date when the Treasury runs out of funds, and though assumed to be a hypothetical date that will always be avoided, it sheds further light on the deeper systemic issues discussed in this chapter and born out of the debt-based monetary architecture we are bound by today.

The introduction of the space value framework, the associated equations of value and return, and their implications for money mechanics offer relevant insights into these systemic challenges. Indeed, Value Easing using PCNs can allow us the opportunity to transform the debt ceiling into a *wealth floor*. While this would be a gradual transformation, it can begin the balanced rehabilitation of an entirely debt-based public expenditure architecture (Papazian 2022).

Furthermore, along with addressing the systemic challenges discussed in the previous sections, such a transformation can provide us with the blueprints of the instruments we can engineer to finance the many trillions of dollars needed to transform our global energy infrastructure and the associated value chains in business and industry.

5.5   CONCLUSION

How can our sustainability frameworks, standards, and scores conveniently omit money creation and the deeper architectural shortcomings of debt-based money? How can sustainability treat the structures and logic of money creation as exogenous to the many challenges we face today on a planetary level? How can money, its value, and the equations that govern its creation, allocation, and deployment be outside the scope of sustainability in finance?

In a transformed financial value framework where space and our responsibility for impact are integrated into our equations of value and return, just like public and private investors, money creators are also required to respect the space value of money principle and ensure that the process of money creation avoids negative impacts across all layers of space and optimises positive planetary, human, and economic impacts.

This chapter explored our current debt-based monetary architecture and identified a number of key limitations of debt-based money, i.e., calendar time, monetary gravity, and monetary hunger. These challenges or features of debt-based money limit and undermine human productivity in space and act like a calendar time linked leash and whip that trigger and often justify the senseless consumption of our planet to meet debt obligations created through our own self-inflicted monetary system. Moreover, our debt-based monetary architecture has legally enforceable primacy over our ecosystem, and our voluntary optional frameworks, standards, and scores of sustainability seem to be unconcerned with this fundamental architectural issue.

The space value framework can usher in a new era in financial and monetary engineering that can facilitate the funding of the many evolutionary challenges we face. The chapter proposed a simple transactional approach that introduces space value creation as an alternative logic to money creation, i.e., Value Easing through Public Capitalisation Notes. This allows us the opportunity to balance an entirely debt-based system, and transform the US debt ceiling into a wealth floor.

And if for any reason one were to doubt the plausibility of such a transformation, suffice it to remember that if the Bank of England, the Federal Reserve, and the European Central Bank can invent and inject money through toxic instruments when needed, they sure can do the same through instruments that have a high positive space impact and can help us address our numerous social, economic, and environmental challenges.

References

Austin, A. 2015. The Debt Limit: History and Recent Increases. Congressional Research Service. https://sgp.fas.org/crs/misc/RL31967.pdf. Accessed 2 February 2020.

Bank of England. 2020. What Are Cryptoassets (Cryptocurrencies)? Bank of England, England. https://www.bankofengland.co.uk/knowledgebank/what-are-cryptocurrencies. Accessed 2 February 2022.

Bank of England. 2021. Bank of England Annual Report and Accounts 20/21. Bank of England. https://www.bankofengland.co.uk/-/media/boe/files/annual-report/2021/boe-2021.pdf?la=en&hash=965204F6565CB8CAD2 9A86E595CB7F02E8A54E07#page=97. Accessed 2 February 2022.

Bank of England. 2023a. Bank of England Annual Report and Accounts 21/22. Bank of England. https://www.bankofengland.co.uk/-/media/boe/files/annual-report/2023/boe-2023.pdf. Accessed 6 July 2023.

Bank of England. 2023b. What is CBDC? https://www.bankofengland.co.uk/explainers/what-is-a-central-bank-digital-currency. Accessed 1 March 2023.

Bernanke, S.B. 2009. The Crisis and the Policy Response. Federal Reserve Board of Directors. Speech at London School of Economics. https://www.federalreserve.gov/newsevents/speech/bernanke20090113a.htm. Accessed 2 February 2021.

Bitcoin. 2021a. Why Do Bitcoins Have Value? Bitcoin.org. https://bitcoin.org/en/faq#why-do-bitcoins-have-value. Accessed 12 December 2021.

Bitcoin. 2021b. How are Bitcoins Created? Bitcoin.org. https://bitcoin.org/en/faq#how-are-bitcoins-created. Accessed 12 December 2021.

Bitcoin. 2021c. How Does Bitcoin Mining Work? Bitcoin.org. https://bitcoin.org/en/faq#how-does-bitcoin-mining-work. Accessed 12 December 2021.

Blanqué, P. 2021. Money and Its Velocity Matter: The Great Comeback of the Quan-tity Equation of Money in an Era of Regime Shift. Amundi Asset management. Discussion Paper 52. https://research-center.amundi.com/article/money-and-its-velocity-matter-great-comeback-quantity-equation-money-era-regime-shift. Accessed 2 March 2022.

Britannica. 2023a. Time Zones. Encyclopedia Britannica. https://www.britannica.com/science/time-zone. Accessed June 2023.

Britannica. 2023b. Meridian Geography. Encyclopaedia Britannica. https://www.britannica.com/science/meridian-geography. Accessed June 2023.

CCAF. 2023. Cambridge Bitcoin Electricity Consumption Index. Cambridge Centre for Alternative Finance, Cambridge. https://ccaf.io/cbeci/index. Accessed 20 March 2023.

CoinMarketCap. 2023. All Cryptocurrencies. CoinMarketCap. https://coinmarketcap.com/. Accessed July 2023.

De Vries, A., and C. Stoll. 2021. Bitcoin's Growing E-waste Problem. Resources Conservation and Recycling. https://doi.org/10.1016/j.rescon rec.2021.105901. Accessed 2 February 2022.

Economist. 2022. The Charm of Cryptocurrencies for White Supremacists. *The Economist.* https://www.economist.com/united-states/2022/02/05/ the-charm-of-cryptocurrencies-for-white-supremacists. Accessed 22 February 2022

Edelberg, W., and L. Sheiner. 2023. How Worried Should We be If the Debt Ceiling Isn't Lifted? Brookings. https://www.brookings.edu/articles/how-worried-should-we-be-if-the-debt-ceiling-isnt-lifted/. Accessed 12 July 2023.

FED. 2022. Total Assets of the Federal Reserve. The Federal Reserve. https:// www.federalreserve.gov/monetarypolicy/bst_recenttrends.htm. Accessed 2 August 2023.

Greene, B. 2004. *The Fabric of the Cosmos: Space, Time, and the Texture of Reality.* London: Penguin Books.

HSBC-BCG. 2021. Delivering Net Zero Supply Chains: The Multi-Trillion Dollar Key to Beat Climate Change. Boston Consulting Group and HSBC. https://www.hsbc.com/-/files/hsbc/news-and-insight/2021/ pdf/211026-delivering-net-zero-supply-chains.pdf?download=1. Accessed 2 February 2022.

IMF. 2023. 3 Global Currency Reserves. International Monetary Fund. https:// data.imf.org/?sk=e6a5f467-c14b-4aa8-9f6d-5a09ec4e62a4. Accessed 31 July 2023.

McKinsey GI. 2022. *The Net-zero Transition What It Would Cost, What It Could Bring.* McKinsey Global Institute. https://www.mckinsey.com/~/ media/mckinsey/business%20functions/sustainability/our%20insights/the% 20net%20zero%20transition%20what%20it%20would%20cost%20what%20it% 20could%20bring/the%20net-zero%20transition-report-january-2022-final. pdf. Accessed 14 February 2022.

McLeay, M., A. Radia, and R. Thomas. 2014a. Money in the Modern Economy: An Introduction. Bank of England. Quarterly Bulletin. https://www.bankof england.co.uk/-/media/boe/files/quarterly-bulletin/2014/money-in-the-modern-economy-an-introduction.pdf. Accessed 6 June 2020.

McLeay, M., A. Radia, and R. Thomas. 2014b. Money Creation in the Modern Economy. Bank of England. Quarterly Bulletin. https://www.bankofengland. co.uk/-/media/boe/files/quarterly-bulletin/2014/money-creation-in-the-modern-economy. Accessed 6 June 2020.

NASA. 2018. Parker Solar Probe Becomes Fastest Ever Spacecraft. NASA. https://blogs.nasa.gov/parkersolarprobe/2018/10/29/parker-solar-probe-becomes-fastest-ever-spacecraft/. Accessed 2 February 2022.

Papazian, Armen. 2022. *The Space Value of Money: Rethinking Finance Beyond Risk and Time*. New York: Palgrave Macmillan. https://doi.org/10.1057/978-1-137-59489-1.

Prasad, E. 2021. Five Myths About Cryptocurrency. *The Washington Post*. https://www.washingtonpost.com/outlook/five-myths/cryptocurrency-yths-bitcoin-dogecoin-musk/2021/05/20/1f3f6c28-b8ad-11eb-96b9-e949d5397de9_story.html. Accessed 02 February 2022.

Rovelli, C. 2018. *The Order of Time*. New York: Riverhead Books.

Smolin, L. 2006. *The Trouble with Physics*. London: Penguin Books.

Statista. 2022. Total Outstanding Public and Private Debt Across All Sectors in the United States from 2000 to 2022. Statista. https://www.statista.com/statistics/1083150/total-us-debt-across-all-sectors/#:~:text=Between%202000%20and%202021%2C%20the,to%2088.2%20trillion%20U.S.%20dollars. Accessed 10 July 2023.

US Treasury. 2023. Debt Limit. US Treasury, USA. https://home.treasury.gov/policy-issues/financial-markets-financial-institutions-and-fiscal-service/debt-limit. Accessed 12 July 2023.

White House. 2023. The Potential Economic Impacts of Various Debt Ceiling Scenarios. The White House. https://www.whitehouse.gov/cea/written-materials/2023/05/03/debt-ceiling-scenarios/. Accessed 23 July 2023.

Withers, W.J.C. 2017. *Zero Degrees: Geographies of the Prime Meridian*. Cambridge: Harvard University Press.

CHAPTER 6

Conclusion

Abstract While the standards and frameworks of sustainability are critically important for a net-zero sustainable world economy, they fall short of transforming our spaceless analytical framework and equations which serve the mortal risk-averse return-maximising investor. The most commonly used tools in the field are content with adjustments to variables in our existing models. Furthermore, the logic and principles that govern the creation of money, although central to the effective conceptualisation and operationalisation of sustainability in finance, are left out of the debate. The space value of money principle and ensuing equations address the above challenges and provide the blueprints for the instruments that could be used to fund the transition. To secure the sustainability and future expansion of human productivity across time and space, planet and humanity must be made equal stakeholders in our equations of value and return.

Keywords Sustainability · Financial mathematics · Money · Value · Risk · Time · Space · Impact

JEL Classification E00 · E58 · G00 · G30 · Q51

Our financial value framework, through the omission of space, as analytical dimension and our physical context stretching from subatomic to interstellar space and every layer in between and beyond, has abstracted away our responsibility for impact in and on space. Through an entirely risk and time-based conceptualisation of money and its valuation, our core equations can be considered directly and indirectly complicit in causing the climate, environmental, and socioeconomic crises we face today (IPCC 2013, 2018, 2021, 2022, 2023; IPBES 2019; Papazian 2022). To reinvent human productivity, we must rethink the financial value framework that has to date tolerated and absolved the littering of our air, rivers, oceans, land, and even outer space.

Indeed, our mathematics of value and return is built in a risktime universe, without space, without context, and without responsibility for space impact. The two principles of value upon which our value and return equations are based, i.e., risk and return and time value of money, discriminate against our own evolutionary investments due to their internal biases against highly risky and very distant cash flows. The mortal risk-averse return-maximising investor and her/his/their tastes and preferences reign supreme in our models.

As the climate crisis and the associated disruptions raise awareness and validate the many and diverse efforts to operationalise sustainability in finance, we are still unable to shift trajectory—from fossil fuel subsidies and investments to exploration licences, the standards, frameworks, scores, and many alliances are yet to deliver the radical transformation we need for effective change of trajectory.

While the standards and frameworks of sustainability are important and critical for our sustainable future, they fall short of transforming our spaceless equations in finance, and they are themselves heavily biased towards risk. Indeed, the most commonly used tools in the field, i.e., ESG ratings and integration, are content with adjustments to variables in our existing models. Furthermore, our sustainability reporting standards and frameworks cannot transform how the reported information is interpreted and used. To achieve effective change, we must reform our financial value framework and rethink the financial mathematics of value and return that has shaped and defined our markets and investments for the last many decades.

Moreover, the logic, principles, and equations that govern the creation, allocation, and deployment of money seem to be left out of the sustainable finance debate, as if they are not the key to any true and effective conceptualisation and operationalisation of sustainability in finance.

How can we reinvent human productivity for Net Zero without rethinking money and its value? Sustainability must be achieved at the level of our monetary architecture as well as our products, businesses, and instruments, public or private. Moreover, the legally enforceable debt-based monetary architecture cannot continue to have primacy over our only ecosystem and the voluntary standards and frameworks that aim to secure its health and sustainability. Indeed, if investors must make responsible and sustainable choices, so must money creators, whether central or commercial banks.

A revised financial value framework involves a change in the very principles that define the value of money. It also involves the introduction of new equations where the space impact of cash flows is considered integral to the value of cash flows. The discounting of future expected cash flows must be accompanied by the compounding of their space impact into the future. The space impact it would take to achieve our future expected cash flows cannot be omitted and/or abstracted away, and planet and humanity must be made equal stakeholders in our models.

A Net Zero sustainable world economy requires a transformed weave of our value chains, our energy systems, productive infrastructures, and institutional architectures. To meet this challenge, which is nothing short of an evolutionary leap, we must hardwire sustainability into the equations that underpin billions of financial and monetary decisions on this planet. Such a transformation can also lead us to the blueprints of the instruments and transactions that could facilitate improvements in money mechanics—a necessary step to create and deploy the funding of the many trillions of dollars we need for the global transition to a healthier and more sustainable world economy.

The space value of money and the transformations it ushers in may very well be one plausible avenue that can lead to a transformed value framework that could, in turn, address the above challenges and secure the sustainability as well as future expansion of human productivity across time and space.

References

IPBES. 2019. The Global Assessment Report on Biodiversity and Ecosystem Services. Intergovernmental Science-Policy Platform on Biodiversity and Ecosystem Services. https://ipbes.net/system/files/2021-06/2020%20IPBES%20GLOBAL%20REPORT%28FIRST%20PART%29_V3_SINGLE.pdf Accessed 02 February 2021.

IPCC. 2013. Climate Change 2013: The Physical Science Basis. Summary for Policymakers. Intergovernmental Panel on Climate Change. https://www.ipcc.ch/site/assets/uploads/2018/03/WG1AR5_SummaryVolume_FINAL.pdf. Accessed 02 February 2021.

IPCC. 2018. Summary for Policymakers. In Global Warming of 1.5°C. IPCC. https://www.ipcc.ch/site/assets/uploads/sites/2/2018/07/SR15_SPM_High_Res.pdf. Accessed 12 December 2020.

IPCC. 2021. Climate Change 2021: The Physical Science Basis. https://www.ipcc.ch/report/ar6/wg1/downloads/report/IPCC_AR6_WGI_SPM_final.pdf. Accessed 02 February 2022.

IPCC, 2022. Climate Change 2022: Impacts, Adaptation and Vulnerability. Summary for Policymakers. Intergovernmental Panel on Climate Change. https://report.ipcc.ch/ar6wg2/pdf/IPCC_AR6_WGII_SummaryForPolicymakers.pdf. Accessed 28 February 2022.

IPCC. 2023. Synthesis Report of the IPCC 6th Assessment Report (AR6): Summary for Policymakers. Intergovernmental Panel on Climate Change. https://report.ipcc.ch/ar6syr/pdf/IPCC_AR6_SYR_SPM.pdf. Accessed 22 March 2023.

Papazian, Armen. 2022. *The Space Value of Money: Rethinking Finance Beyond Risk and Time.* New York: Palgrave Macmillan. https://doi.org/10.1057/978-1-137-59489-1.

INDEX

Milton Keynes UK
Ingram Content Group UK Ltd.
UKHW021508171123
432744UK00002B/24